THE NEW HERDSMAN'S BOOK

THE NEW HERDSMAN'S BOOK

MALCOLM STANSFIELD

FARMING PRESS LIMITED
Wharfedale Road, Ipswich, Suffolk IP1 4LG.

First published 1983

ISBN 0 85236 133 5

Copyright © Farming Press Ltd, 1983

Phototypeset by Galleon Photosetting, Ipswich
Printed and bound in Great Britain at the Camelot Press Ltd, Southampton

CONTENTS

LIST OF PLATES

LIST OF FIGURES

PREFACE

THIS BOOK appears, not because of any ambitions to become an author, but due to a concern that people working with cattle should do a good job, make the most of their changing situations and obtain maximum satisfaction from their work. Even my school reports would endorse this as an appropriate objective, especially those from my English Master who stated with some regularity, 'If this pupil took half as much interest in his studies as he does in his farming he would go far'! He would smile today if he only knew how many drafts have been necessary to produce this text.

My interest in cattle was aroused at an early age on the local dairy farm of Brewis Cant, a friend of the family. He typified the first-class stockmen of the Yorkshire Dales but perhaps was exceptional in his patience, understanding and ability to get the most out of youngsters, as well as cows. I was taught the correct ways of undertaking husbandry tasks: that the cows come first but that every farm resource is important and needs care, whether it be a field of baled hay or the yard broom. The factors which determine success in dairying were clearly demonstrated, such as: good feeding, regular breeding and prompt attention to a sick animal. Most importantly, I was taught to *think* before acting and to complete every task to the best of my ability, especially when cattle and other people were involved.

This book appears, therefore, because I consider that so many of those factors continue to be critical to success. The modern herdsman may well be operating with cattle of higher potential and with a wide range of sophisticated aids, but he or she is still working with the same basic resources; and so as well as a knowledge of modern technology, they need the right attitude to the job.

It is a great privilege to have my name on a publication which in many ways follows Kenneth Russell's original *Herdsman's Book*. I first met Ken when I was visiting Askham Bryan as an undergraduate student and in later years I heard him speak to farmers on many occasions. He was a truly great man and his words on stockmanship were 'music to my ears'. On the day he arrived in the farmyard at Sonning for a walk through 'my' herd and a discussion on the management of large dairies, I thought I had then achieved all that was possible in dairying.

Like his books, this one is written in the hope that it will be of help to practical herdsmen, to their employers, as well as to all students of dairying. Like Ken, too, I admire good cows and good herdsmen—and the one deserves the other.

I am grateful to many colleagues and friends in the industry for their help, advice and supply of information. My very special thanks are due to Tony Giles and to Bill Milligan for their invaluable encouragement and also for reading and suggesting amendments to major parts of the manuscript.

I would also like to thank Mrs Frances Green and Mrs Sheila Smith for typing the several drafts and for their general help in the preparation.

To the many local farmers and herdsmen and especially to the staff of Sonning Farm who so willingly co-operated in the production of photographs, I am also most grateful. My thanks are also due to Ian Maclean who took the majority of the pictures, patiently protecting his equipment from slurry and the licking of friendly cows. The remainder were kindly supplied by BOCM Silcock Ltd., *Dairy Farmer*, Tim Bryce, Tony Quick and my son John.

Finally, my special thanks to my long-suffering wife, Mary, who began to believe it would never be finished—and to Roger Smith of Farming Press for his constant help and advice in order to ensure that it was.

INTRODUCTION

WITH THE continuing trend towards mechanisation and automation of livestock farms some may wonder if the days of the stockman* are numbered. It is a fact that on many dairy farms electronically controlled equipment has replaced the herdsman for such tasks as concentrate feeding and cluster removal. On a few units, automation of milk recording, teat disinfection and cow weighing is in operation, with results usually more accurate than when undertaken by stockmen. In the opinion of this writer, however, it will be a remarkable day when the dairy robot is able to cope with a difficult calving or to persuade a dozy, new-born calf to suckle its dam!

In the future, many aspects of the herdsman's job will no doubt continue to change. He may further reduce his physical work and spend more time supervising equipment, but he will retain, and even expand, the time spent on the vital task of caring for the welfare of his cattle.

Change is not a new experience for most herdsmen, at least for those working with cattle during the last fifty years. The average dairy herd has more than doubled in size, breeds of both dairy and beef cattle have come and gone and machinery has progressively replaced much of the hand work involved in feeding, milking and manure handling. Specialisation has become a feature of many farms, with herds being increased and systems intensified to cover the rising background costs of such items as labour and machinery.

Not all cattle enterprises have changed quite so dramatically. Although very few animals are now hand milked, many smaller farms still operate with cowshed housing and use minimal mechanisation. The change in the role of beef herdsmen has in general been less marked than with dairying especially for those involved

---

* or woman. It would be invidious to suggest that one sex was better at caring for stock than the other. A single gender will be used to avoid unnecessary repetition.

1

1. Dispersing from a modern milking parlour. The front cow is entering an electronically controlled teat disinfection unit, and the second walking over an automatic weigher.

2. A large herd ready for milking. This 160-cow herd is operated by one herdsman, aided by numerous labour saving devices. The centrally mounted gate in the rotary collecting yard moves around behind the cows, encouraging them to enter the parlour.

with suckler herds. Many such herds are carried on holdings where the climatic and physical environment limits the economic intensification of the enterprise. Others are integrated as low-cost units utilising by-products on arable farms. However, on many beef-fattening units stockmen are involved in technology as high as that in dairying with sophisticated feeding and management systems.

The detailed tasks that make up the job of each stockman do therefore vary considerably from farm to farm but there are some which are common to all situations. An attempt has been made in the following chapters to discuss all the most important tasks and considerations so that the book should be of interest to everyone involved with cattle enterprises regardless of their types or sizes.

It is assumed that all herdsmen are keen and dedicated to their cattle, being fully aware of the satisfaction that can be obtained from a job well done. One of the great pleasures of working with stock is to see the results of one's efforts, sometimes almost immediately, as with a cow responding to injection for grass staggers. The job also has its share of frustrations and disappointments—sometimes even more than its share!

The objective of this book is to foster a balanced approach to the job and to encourage herdsmen to think and plan; but above all to carry out timely and appropriate action. Hopefully it will encourage herdsmen to correct their weaknesses (which everyone has) and to build on strengths by developing skills (which everyone also has). Tackling a job with understanding and confidence leads to even greater satisfaction and contributes much to improvements in the physical and financial performance of the enterprise.

Financial aspects of the herd have traditionally been of minor interest to herdsmen but this is a situation which is now rapidly changing. In most high-performance enterprises, farmers and managers involve their herdsmen in many of the business tasks such as setting targets and discussing results, in a way which usually provides a considerable stimulus to motivation.

On the numerous one-man units, many of which are family farms, the herd owner is also stockman so that he has even more demands on his time. Routine tasks such as feeding naturally take priority, but when problems arise such as a sick cow the business aspects and office work just have to wait.

In all herd situations there is an important need for effective communications between herdsmen and the various people who are involved in the operation of the enterprise. With small herds, apart from members of the family, this may involve only the veterinary surgeon, AI inseminator and perhaps a relief milker. In large herds it will involve other herdsmen and often tractor drivers, staff from the farm's other enterprises as well as a wide range of off-farm people such as advisors and representatives.

Where there are large herds with several herdsmen involved a good team spirit is essential to success. Each member needs to enjoy working with other people and to understand the interests and motivations of his colleagues. Sharing the less pleasant aspects of the work such as slurry scraping always helps. There is a considerable challenge for employers and managers in these conditions to set up and control an effective system of operation. Team members need to be selected with suitable personalities for this rather special situation. Researchers have shown that many herdsmen prefer to work alone and to take full responsibility for the day-to-day operation of a herd. This type of herdsman apparently tends to be an introvert and may be keen on such hobbies as gardening. He is not concerned about the long working day which offers few opportunities to get away from the farm. However, herdsmen operating in teams commonly work shifts so that at intervals during the rota they have useful blocks of time which can be spent with the family (at meal times) or participating in leisure activities off the farm.

It appears, therefore, that despite the arrival of modern technology on the cattle farm there is still an important job to be done by herdsmen. The successful ones will increasingly be those with a wide range of skills, a detailed knowledge of what 'makes a cow tick', but with the ability to carry out the right job at the right time. They will also have an increasingly important role in the general management of cattle enterprises by being 'spokesmen' for the cattle in making appropriate representations when decisions need to be taken about such items as building design, seed mixtures and diet formulation.

It is the author's hope that this book will help stockmen to undertake this changing role with confidence.

Chapter 1

GENERAL STOCKMANSHIP

THERE IS no doubt some truth in the saying that a good stockman is born and not trained, as many of his essential qualities such as patience, reliability and a sense of caring are very much part of his character. However, other equally essential qualities such as his skills of observation, handling of stock and the undertaking of his many husbandry tasks are gained by experience and training.

Stockmanship is not easy to define. It is well recognised that the person with such talents has 'a way with stock'. There exists a special relationship between man and animal, each understands the other and they even appear to be able to communicate with each other. Watching a skilled herdsman amongst his cattle, one can see stockmanship in action. He operates with confidence, speaks to the animals in a reassuring tone of voice and often touches them so that they show no signs of fear and seldom back away.

The results of his sense of caring can be seen in the general health of the stock, and their environment: the bloom of the coat, clean, dry cubicle beds, the absence of overgrown feet.

Traditionally, the herdsman had to know every animal in the herd so as to be able to manage each one as an individual. The situation in many of today's larger herds, with several herdsmen involved and with considerable automation, is somewhat different. Nevertheless, any animal seen to be behaving in an abnormal or unexpected way has to be immediately identified and the records checked before any subsequent action can be taken.

Observation

Careful and critical observation, not only of individual animals but also of the group as a whole, is therefore an essential feature of good stockmanship. It involves most of the senses—sight, touch, smell, hearing—but rarely taste! Careful observation can take quite a lot of

5

3. Young herdsmen and women taking part in a training session. The training officer is demonstrating the correct method of dosing a young heifer.

4. A demonstration of stockmanship. The good herdsman operates with confidence, speaks to his animals in a reassuring tone of voice and often touches them, so that they show no signs of fear and seldom back away.

energy so that at times of heavy physical work there is a danger of becoming over-tired and observing less accurately.

Skilled observation comes with experience but it can be greatly enhanced by training, especially by working alongside a skilled stockman who at every opportunity points out to the trainee any irregularities and confirms the normal situations.

It is not only the animals which have to be observed, attention must also be given to the numerous aspects of the environment which influence animal behaviour and performance. Excessively wet bedding may for example indicate a fracture in an underground water pipe or a blocked drain rather than any problem with the straw or with the animals. Attention must even be given to colleagues, especially when one is responsible for *their* work. A person under the weather may not be aware that his performance is slipping but if he is not fully alert he is more likely to be involved in an accident.

Animal Behaviour

A herdsman will find it easier to observe and handle the animal effectively if he is aware of the normal behavioural pattern of a herd of cattle. A number of factors influence behaviour, one of the most obvious ones being the oestrous cycle in cows and mature heifers—which will be discussed in Chapter 11.

The social ranking within a herd has a marked effect upon behaviour. Some animals are more aggressive than others and there is an aggressive ranking or 'bunt order' within a group, with the boss animals at the top and timid ones at the bottom. The larger, older cows in a milking group tend to be of high rank and first-lactation heifers and recently calved animals of lower rank. This factor is particularly important in the management of a herd which is loose-housed or self-fed silage. Boss cows will stand at strategic places such as the silage face, keeping subordinate ones away so that they fail to get adequate feed. Tombstone barriers or feed fences fitted with diagonal rails reduce such bullying as boss animals are prevented from moving quickly back from the feeding face. Subdivision of a large group into sub-groups is a management practice which also reduces bullying. A milking cow in a group of no more than eighty seems to be able to readily recognise the ones higher in the 'bunt order' and so moves to one side to let them pass. In groups with over a hundred cows such recognition is not possible and more bullying occurs. The behavioural problems in moving cows from one group to another are discussed in Chapter 5.

Herd Movement

Cattle moving of their own volition tend to string out in a line with natural leaders at the front. When a herd is driven, for example along a trackway, the ones at the back are under some pressure so that the herd forms more of a triangular shape. Stress is caused if cattle are rushed at a pace faster than the normal level of the slower members of the group. It is easy to damage feet and udders and cause poor let-down if the cows are rushed just before milking.

Ideally, a milking herd comes to the gate when called, avoiding the need to drive them from a field or yard or to use a dog for collection. Few dogs are sufficiently gentle when working with cattle, and have a tendency to nip heels and cause too much disturbance. Gathering cattle from a large field can be time-consuming and frustrating but the author is not impressed by herdsmen who dash around the field on a tractor trying to out-perform a fast dog!

Separating one animal out of a herd, such as a down-calver from a group of dry cows, is another operation requiring considerable skill on the part of the herdsman. A stick is a useful aid on such occasions, not for hitting the animals but to show them that you do mean business, and it should always be kept down because a stick held high seems to alarm the stock. If the animal you are trying to separate is nervous it may be necessary to take more than one out of the group so as to accompany her to the new location.

An inexperienced handler spends too much time running after cattle trying to recover a situation from which they have escaped. The skill comes in avoiding a breakaway by watching the group carefully, working and moving quickly but without too much noise. Cattle move easier if they have only one obvious way to go. Loading into a cattle truck, for example, is made much simpler if the vehicle can be reversed up to a race which has high sides so that there is no chance of the animals jumping over. A little straw on the ramp usually helps reduce the fear of loading, as does allowing them time to smell their way into the 'unknown'.

Teaching to Lead

Show cattle and young bulls need to be halter trained and this is best started when they are young calves. At this age they can easily be tied up with light halters to a ring in a wall, so that they become used to being restrained.

Once accustomed to a halter they continue to respond to safe handling in this way as adult cattle. When fitting a halter it is best to

5. Checking a group of cows on an ad-lib diet. Careful and critical observation is an essential feature of good stockmanship. A check during feeding is usually the best time; this is more difficult to achieve on an ad-lib system where only a proportion of the herd is eating at any one time.

6. First-lactation heifers on a self-feed silage system. In larger herds and where building layout permits, it is advantageous to have a separate heifer group so as to reduce bullying by older and larger cows. This enables increased food intake to be achieved, providing nutrients for maintenance, milk production and adequate liveweight gain.

7. Moving towards afternoon milking. Stress is created if cattle are rushed along tracks; it can cause damage to feet and to udders.

8. Halter training a young heifer. Show cattle and young bulls need to be halter trained and this is best started at the calf stage.

put the poll piece over the ears first, before tightening the loop around the nose. Bulls of course need to be handled from yearlings with a pole as described in Chapter 11.

Casting

This method of handling cattle is now rarely used. Even when operations have to be carried out on the feet a modern crush is preferable for restraining the animal. However, a good stockman

Figure 1. Reuff's method of casting cattle.

should know how to cast cattle using the Reuff method. The animal is first haltered so that the head can be controlled, preferably by an experienced person. A long rope is used, first placing a loop round the neck and securing it above the withers. The long end of the rope is then taken along the back forming a half-hitch around the chest just behind the front legs and another around the abdomen in front of the udder in a cow. If one or two people then pull steadily on the rope the animal should collapse to the floor. The knot used in Reuff's casting is a quick-release bowline.

Using the Crush

This is the most satisfactory method of restraining cattle in which stocks hold the head firmly and a bar is placed behind the rump. Most animals tend to be nervous when they enter the crush so that quiet but firm handling is required. The approach race should be strong, safe and free from any protrusions, which could injure the cattle. The floor of the race, especially if concreted, should be kept

9. A modern hydraulically-operated crush. The mechanism requires regular servicing and the wooden floor must be kept clean and in good repair.

10. A cow tipped on its side in the hydraulically-operated crush and immobilised ready for foot treatment.

clean so as to minimise slipping. The floor of the crush is often constructed of timber, especially if the equipment doubles-up as a weighbridge; this certainly needs to be kept clean and in a good state of repair. Moving parts such as the gate hinges and any weighing mechanism should be oiled frequently.

It is essential when using the crush to be alert and avoid being trapped, especially when releasing an animal from the front gate.

Restraining Without a Crush

It is often necessary to restrain a beast in a location where there is no access to a crush. If an assistant is available it may be possible to use a lightweight gate to trap the animal against a pen wall; then with thumb and first finger take a firm grip of the nostrils. Fingernails obviously need to be short to avoid damaging the tissue of the nostril, and if the animal is mature considerable strength is required!

Drenching

This is a job best carried out using a crush although with the occasional animal the nostril grip will be adequate. The medicine to be administered should be placed in a long-necked plastic bottle. If you are right-handed, you should stand on the right-hand side of the animal restraining its head with the left hand. The bottle is placed in the side of the mouth above the tongue and just behind the incisor teeth. The fingers of the left hand are placed in the same gap at the opposite side so encouraging the mouth to open. Care has to be taken to allow swallowing, giving a pause between each dose.

Injections

This is a method being increasingly used to administer drugs and one in which the herdsman needs to be particularly skilled. Care and attention to cleanliness, knowledge of the correct type and site of injections and selection of appropriate needle are essential. Initial supervision by a vet or training officer is advisable before undertaking the task alone.

The Subcutaneous (under the skin) method is commonly used and in cattle is ideally carried out at the base of the neck, in front of the shoulder. The site should first be swabbed with skin disinfectant, a pinch of skin taken and the needle carefully inserted. After injecting the medicine and withdrawing the needle a gentle massage of the area will help disperse the material from the site.

11. Drenching a cow. If a large group has to be dosed, it is a job best carried out in the crush, but when only the occasional animal requires treatment catching with the nostrils is adequate.

12. A subcutaneous injection for the control of foul-in-the-foot.

Intramuscular (into the muscle) injections are usually carried out into the buttocks. The needle should be detached from the syringe and held firmly in the finger and thumb. A few firm taps on to the site (previously disinfected) with the back of the fist prepares the animal so that when the needle is plunged into the muscle it is not frightened.

Intravenous (into a vein) injections do need rather more skill and training. The jugular vein along the side of the neck is a good site to select. A point along the neck should be clipped and swabbed with spirit. With mature cows and bulls it is necessary to use a rope noose which, when tightened, distends the vein. A short, sharp, bevelled needle can then be inserted into the vein, syringe attached and medicine pushed steadily into the bloodstream while the rope is released. After you remove the needle hold the thumb firmly over the site to prevent any loss of blood.

Inter-Uterine (into the uterus) injections are frequently required to control metritis. The insertion of a catheter is usually a job for the vet but when undertaken by herdsmen needs to be done with much care to avoid tissue damage.

DAILY ROUTINE

An example of a herdsman's daily routine, during the winter months in a one-man unit involving parlour milking, cubicle housing and self-feeding of silage.

Task

1. On arrival, check the down-calving cows. If an animal needs assistance allow 5–10 minutes to deliver the calf, otherwise call for help to avoid undue delay in milking (especially if the milk tanker comes early).
2. Final preparation of the milking equipment including check on bulk tank temperature (main preparation previous evening).
3. Bring cows to collecting yard checking on any cows in oestrus.
4. Switch on the vacuum pump, check the gauge and undertake milking; this will include concentrate feeding.
5. Wash parlour and sterilise milking equipment.
6. Check breeding/health records of cows diverted for treatment. Phone AI Centre to order inseminations.

Breakfast

7. Scrape slurry from cubicle passageways. Check water troughs. Top-up bedding in cubicles (alternate mornings).
8. Feed and bed-up dry cows.
9. Provide a mid-morning manger feed to milking cows (sugar beet pulp, concentrates). Observe cows while they are eating, then scrape slurry from self-feed area. Adjust feed barrier or fence.
10. Assist vet or AI inseminator. Treat other problem cows, e.g. lame ones.
11. Feed calves on milk substitute (using once-a-day system).
12. Late morning check for cows in oestrus (10–15 minutes).

Lunch

13. Check once again the down-calvers.
14. Complete the washing/disinfection of the bulk milk tank following collection and activation of automatic washer by tanker driver. Prepare parlour for milking, replacing any damaged rubbers.
15. Undertake any essential maintenance such as a broken cubicle rail.
16. Complete the records for the day, i.e. AI, calvings, vet, mastitis. Ear-tag calves and sketch pedigree cards.
17. Bring cows to collecting yard then scrape slurry from cubicle passageways.
18. Afternoon milking, concentrate feeding.
19. Check down-calvers, dry cows and calves.

Tea/evening meal

20. Late-night check of all stock, especially oestrus detection (20–30 mins.)

Sleep well!

ANNUAL ROUTINE

An example of the routine tasks undertaken by a herdsman during the year. The example is from an autumn-calving herd so that the Production Year begins in July.

July   **Summer holidays**—suitably refreshed for the new production year.
2nd-cut silage. Start drying-off early season calvers. Dose yearling heifers against intestinal parasites.

Aug.   Receive annual straw requirement, arrange to have a quantity chemically treated for feeding purposes.
Reseeding of several grass paddocks. 3rd-cut silage.

Sept.   Start of calving season for heifers (22–24 months old). Meeting with feed adviser to plan winter rations. Cull the cows that failed to conceive and have gained weight at grass. Visit Dairy Farming Event to keep abreast of new technology.

Oct.   Start of calving season for cows.
Meeting with AI centre manager to finalise semen requirements for forthcoming breeding season.
Open one silage clamp and slowly introduce winter diets to freshly-calved animals. All cows housed by night.

Nov.   Peak of calving season.
Meeting with veterinary surgeon to finalise plans for breeding season.
All cows and youngstock housed for winter months.
Attend local Herdsmen's Club meeting on 'Breeding the Cow of the Future'.

Dec.   Start of the breeding season. Apply tail paint to all cows not seen in oestrus by 40 days of lactation.
Meeting with herd owner to discuss possible new equipment for next silage making season. (Smithfield Show.)
Arrange for prostaglandin treatment of heifers and double AI.
Make preparations to minimise daily routines over Christmas period.

Jan.   Short break over New Year
Run a beef bull with heifers to cover returns to oestrus.
Meeting with herd owner and costings adviser to discuss
annual performance to date and predicted outcome for
financial year ending 31st March. Consider possible capital
replacements and improvements.

Feb.   Plan grassland use for forthcoming grazing/conservation
season.
Apply 1st nitrogen applications.
Continue pregnancy diagnoses.
Arrange for slurry store to be emptied.
Check that order has been placed for silage additive and
plastic sheeting.

Mar. Check fences and field water supplies.
Cull further cows when prices are high (especially if winter
feed stocks in short supply)
Arrange for fields due to be cut for silage to be flat rolled.

Apr.   Turn out older heifers, then cows by day.
Slowly adjust diets as grass growth improves.
Vaccinate heifer calves for husk.
Cows out by night by end of month.

Spring break
May   Prepare silage clamps for new season's crop.
Pressure wash and disinfect cubicle house.
Continue to apply fertiliser after each paddock grazed.
Turn out young calves to 'clean' grass.
3rd week—assist with 1st cut silage campaign.

June   Fit fly-control ear tags to in-calf heifers.
Provide data for completion of MAFF June 4th returns.
Carry out essential maintenance of buildings and equipment.
Visit other farms to see progeny of bulls which could be used
in the herd.
Brief relief herdsman ready for forthcoming holidays.

Much satisfaction can be obtained by undertaking these husbandry
tasks in a correct and safe way. Care of equipment is essential. When
the task is completed tools and equipment should be cleaned and

returned to the store and injection needles should be sterilised ready for use on the next occasion.

The reader of this chapter may now appreciate why its author claimed at the outset that there is *some* truth in the saying that good stockmen are born, not trained—but that, equally, some important qualities and skills can be gained by experience and training. More often than not good stockmanship exists precisely because of that kind of combination of natural and acquired skills and the aim of this chapter has been to demonstrate that fact.

Chapter 2

PRINCIPLES OF FEEDING

ANOTHER MAJOR factor affecting the profitability of a cattle enterprise is the provision of feed for the stock which is adequate in both nutritional and economic terms. Carefully controlled feeding experiments have been undertaken to obtain the precise nutritional requirements of all classes of cattle. Costings schemes have been worked out including computer programmes which, when provided with information regarding available feeds and prices, calculate optimum or least-cost rations.

Despite all these developments, there continues to be a most important role for herdsmen in modifying such 'standard' figures to meet the day-by-day or even hour-by-hour needs of his stock. An animal which has been 'off-colour' may need to receive a special diet for several days before full recovery takes place. In order that the herdsman can undertake the feeding of the herd effectively and with confidence, it is necessary for him to first understand some of the principles or science of feeding.

The Digestive System

Cattle use their food for two main purposes: first to provide the means of maintaining life and normal activity; this is known as the maintenance requirements. Secondly, for the production of milk, meat and the development of the unborn calf. Following digestion, the food provides the body tissues with energy, protein, minerals, vitamins and some water (especially from foods such as grass, kale and silage). It will therefore be useful, at this juncture, to consider briefly how the digestive system does in fact work.

In essence it consists of a long feed tube passing from the mouth to the anus as shown in Figure 2. The food is broken down in a series of processes; firstly mechanical action, primarily chewing, secondly microbial fermentation particularly in the rumen and, thirdly, by

20

biochemical digestion involving enzymes which takes place at various parts of the gut. The nutrients or products of digestion pass through the wall of the digestive tract into the blood stream for transport to other organs such as the udder for the production of milk.

Food is taken into the mouth by a grasping and tearing action using the lips, teeth and tongue. The rear molar teeth then grind the food before swallowing, as they do again when a bolus of regurgitated food is further chewed during rumination (cudding). From the

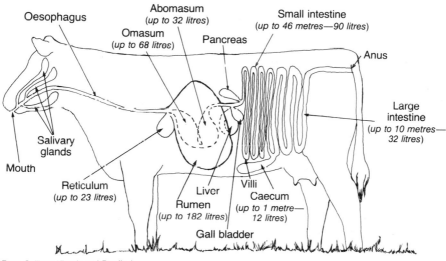

From Cullison (*Feeds and Feeding*).

Figure 2. The digestive tract of the cow.

mouth, food passes down the oesophagus to the first of the four stomachs known as the rumen. It is the largest of the four and is attached to the much smaller reticulum which is located in front of the rumen. Also opening from the rumen and lying to the right is the small round omasum which is followed by the abomasum or true stomach.

The *reticulum* and *rumen* work together as the main fermentation site. The vessel contains millions of microbes which are involved in breaking down the food into a fine liquid mass. Some of the larger pieces of food are returned to the mouth during rumination by reverse action of the muscles. Fermentation is a continual process hence the need for regular feeding to keep the microbes working to the full. Large amounts of saliva are required in fermentation, this is produced from the glands in the mouth (up to 150 litres per day in a high-yielding cow). This provides water to assist the chewing process and salts to buffer the acids formed from the breakdown of carbohydrates. Gases such as carbon dioxide and methane are produced during fermentation and in the healthy animal are removed by belching or during rumination. Bloat becomes a problem in an animal which is unable to remove gases in this way. A regular supply of fibrous food keeps the vessel in good function and helps to prevent bloat.

Most of the carbohydrates (energy-supplying substances) are broken down in the reticulo-rumen to volatile fatty acids such as acetic, propionic and butyric acid. Simple sugars are broken down rapidly but cellulose and hemi-cellulose from the cell walls of plants are degraded much more slowly.

Modification of oils and fats also take place in the reticulo-rumen and some products of fermentation are absorbed into the blood stream but most 'feed' the microbes which then pass down the alimentary tract for subsequent digestion.

The *omasum* has a lining of tissue similar to the pages of a book. Its main function is to remove surplus water from the material before it flows into the abomasum where the main biochemical digestion begins.

The *abomasum* secretes large amounts of acidic gastric juices which kill the microbes and start the process of protein digestion. The material then flows into the U-shaped duodenum receiving enzymes from the pancreas and bile from the gall bladder to neutralise the acidity of the residual gastric juices.

The *small intestine* is extremely long (up to 45 metres). It is the main area of protein absorption and the site for some glucose production from the digestion of carbohydrates.

The *colon* or *large intestine* is not as long as the small intestine but has a bigger diameter. Excess water is removed from the gut contents at this point, so concentrating the faeces. Some micro-organisms are at work in the colon, further fermenting fibre as in the caecum which also acts as a secondary fermentation chamber.

Another major organ involved in food digestion is the *liver*. It acts as a store of energy, deals with ammonia in the blood and converts numerous complicated chemicals into much simpler ones for use by other body organs. The *kidneys* too have an important role in taking waste products from the blood, these being discharged in the form of urine.

It can be seen that apart from eating, defecating and urinating which take place at intervals, digestion is a continuous process. Any

13. Yearling Hereford x Friesian steers eating maize silage at a feed fence. The diagonal rails of the fence deter the dominant animals from backing out to change feeding positions when the concentrate part of the ration is fed.

part that suffers from an upset creates problems to the whole system. Herdsmen therefore have an important role in providing feed, water and the right environment so that efficient digestion takes place.

Feed Requirements

The method of expressing feed requirements in the United Kingdom has since 1975 been based on the Ministry of Agriculture Technical Bulletin No. 33 which replaced the previous Starch Equivalent System. The new system uses the Metabolisable Energy (ME) in the food as a basis for formulating rations. The ME is the energy remaining after the loss in faeces, urine, gases and body heat. The basic unit used to measure the energy is the Megajoule (MJ).

The basis of a rationing system is to meet the energy needs of the animal from the foods provided, but also taking into account the digestible crude protein (DCP) requirements. The requirements of ME and DCP for maintenance are supplied in Table 2.1 and for production of milk in Table 2.2. The ME and DCP considerations in liveweight gain and loss are supplied in Table 2.3.

Dry Matter Intake

The next consideration is an important one, being the amount of food that the animal will eat. Appetite has been shown to vary with the size of the animal, its level of production but also with the

Requirements for Energy and DCP

Table 2.1   Maintenance only

L.Wt. (kg)	ME (MJ/day)	DCP (g/day)
300	36	200
350	40	250
400	45	250
450	49	300
500	54	300
550	59	350
600	63	350
650	68	400

Table 2.2   Production of 1 kg milk

Type of milk	B.F. %	S.N.F. %	ME (MJ/day)	DCP (g/day)
Friesian	3.6	8.7	4.98	60
Ayrshire	3.8	8.9	5.17	65
Channel Island	4.8	9.2	5.93	70

Table 2.3   Live weight change

Gain/loss	MJ (kg/day)	DCP (g/day)
+ 1 kg	34	320
− 1 kg	28	—

Table 2.4   Probable dry matter intake in early lactation (weeks 1−10)

L.Wt. (kg)	Dry	Milk Yield (kg/day) 10	20	30	40	50
350	4.8	8.3	11.8	15.3	18.8	22.3
400	5.4	8.9	12.4	15.9	19.4	22.9
450	6.0	9.5	13.0	16.5	20.1	23.9
500	6.6	10.2	13.7	17.2	20.7	24.2
550	7.3	10.8	14.3	17.8	21.3	24.8
600	7.9	11.4	14.9	18.4	21.9	25.4
650	8.5	12.0	15.5	19.0	22.6	26.1

digestibility of the ration, its palatability and the manner in which the food is presented. A guide to dry matter intake (DMI) of cows at different levels of yield is supplied in Table 2.4.

Intake is actually dependent on the speed at which food passes through the digestive tract. Low fibre, laxative types of diet such as spring grass pass through so quickly that digestion is inefficient and the animal may lack sufficient nutrients. High-fibre diets on the other hand, can be so indigestible that again the requirements of energy and protein fail to be met.

Calculating a Ration

In this example a cow of 550 kg liveweight is producing 30 kg milk at 3.6% B.F. and 8.7% S.N.F. She is in early lactation and is losing weight at 0.5 kg per day. (Composition of feeds, Table 2.5.)

REQUIREMENTS

	ME (MJ)	DCP (g/day)
Maintenance (550 kg)	59	350
Milk production (30 kg)	149	1800
Liveweight loss (0.5 kg)	(14)	0
Total	194	2150
Probable Dry Matter Intake	17.8 kg	

RATION

	Fresh Wt. (kg)	DM (kg)	ME (MJ)	DCP (g)
Grass silage (high D)	40	10	103	1050
Wet sugar beet pulp	5	0.9	11	60
Concentrates (standard)	8	6.9	86	1104
	53	17.8	200	2214

This diet adequately meets the energy and DCP requirements of the cow and achieves this within the probable level of dry matter intake.

Protein Degradability

Although, as explained above, the DCP method of calculating the protein requirement of cattle is widely used, a new system is currently being introduced which takes into account the degradability of the protein in the ration. It is a superior system of calculating the requirement, especially for high-yielding cows which

Table 2.5   The composition of some common cattle feeds

Feed	Dry Matter (g/kg)	ME (MJ/kg DM)	DCP (g/kg DM)
Grazing (rotational)	200	12.1	185
Grazing (extensive)	200	10.0	124
Silage (high 'D')	250	10.3	105
Silage (moderate 'D')	250	9.8	102
Hay (high 'D')	850	9.0	58
Hay (moderate 'D')	850	8.4	39
Straw (spring barley)	860	7.3	9
Straw (winter barley)	860	5.8	8
Kale	140	11.2	106
Mangoes	120	12.4	54
Potatoes	210	12.5	47
Barley	860	12.9	82
Oats	860	12.0	84
Maize	860	14.2	105
S.B. pulp (wet)	180	12.7	66
Brewers' grains (wet)	220	10.0	149
High energy dairy compound	860	14.0	180
Standard dairy compound	860	12.5	160

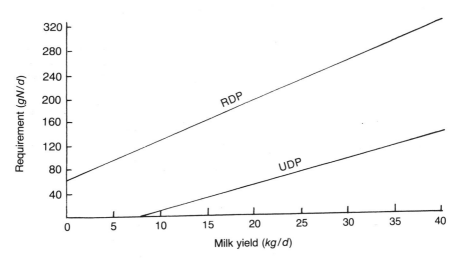

From: Wilson and Brigstock.

Figure 3. Protein requirements related to milk yield.

have been shown to benefit from protein which escapes degradation in the rumen and is absorbed as amino-acids in the hind gut. These types of protein are known as Undegradable (UDP), good sources being fish meal, meat and bone meal and grass silage which has been treated with formaldehyde. Foods with medium degradability are dried grass, soya bean meal and maize silage. Those which are degraded in the rumen (RDP) are used by the micro-organisms, common sources being grass, hay and silage without additive. The rumen micro-organisms can also utilise non-protein nitrogen (NPN) such as urea which has for some time been fed to beef cattle but is now being incorporated into dairy diets.

If sufficient RDP is available, the rate of fermentation of fibrous as well as concentrate rich diets will be reduced. This leads to a reduction in intake, lower energy supply and reduced milk production.

It is therefore important to have the optimum balance of UDP and RDP in the diet. Figure 3 illustrates the increasing requirements for both as milk yield increases.

To obtain the appropriate levels it is necessary to calculate the total energy and energy concentration in the diet for the desired level of yield. Then determine how much total protein is required to match that energy. After calculating how much can be provided by the microbes, the remainder has to be provided in an undegradable form. The feed trade and other advisory organisations have computer programmes which deal with this aspect of ration formulation.

Mineral Requirements

The major mineral requirements for dairy cows are calcium and phosphorus. The requirements for maintenance and production of a 6,000 litres lactation are 50 g calcium and 70 g phosphorus per day. Winter diets can be short of calcium and pasture deficient in phosphorus. Shortages of sodium are rare but salt licks are useful, especially when high rates of potassium fertiliser are used. Magnesium is frequently required, at the rate of 50 g per day, to prevent hypomagnesaemia.

Vitamin Requirements

The only vitamins likely to be in short supply are A and D in the winter months. Vitamin A is stored in body tissue from summer grazing with kale and high-quality silage also being a good source. Vitamin D is provided in well-cured hay and concentrates are often supplemented with synthetic A and D.

Before concluding the principles of feeding and moving in the next two chapters to the practical aspects of feeding in winter and in summer, it is necessary to stress the long-term effects of feeding a dairy cow. The feeding level at one point of the lactation can have a marked effect upon yield and body condition, not only later in that lactation but also in subsequent lactations. This is well illustrated by

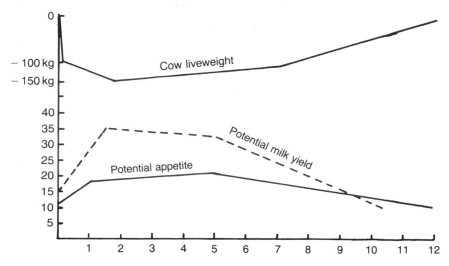

Figure 4. Relationship between milk yield, appetite and liveweight.

good feeding in late lactation which increases body condition at calving so that a higher peak yield is achieved as well as a higher total lactation yield.

Figure 4 demonstrates the relationship between milk yield, body weight change and appetite. The shape of the lactation curve is affected by genetic potential but also by feeding and management. The peak yield is critical, because for each litre obtained at peak the overall lactation increases by 200 litres. Rate of decline after peak is normally 2.5% per week so that the yield around the 100th day of lactation is often a better indicator of total yield than is peak yield.

Appetite is lowest at calving and it increases to a peak around the fourth month of lactation before declining as the developing foetus takes up body space. The energy deficit from the feed in early lactation can be compensated by the mobilisation of body fat in the first two months after calving. It is therefore necessary to ensure that cows are in good body condition at calving. The regaining of weight before the subsequent calving is more efficiently undertaken while

14. Milking cows retained in self-locking yokes to allow individual rationing of concentrates according to yield. Note the large plastic ear-tag used for identification.

the cow is still milking rather than when dry. Overfeeding during the dry period can also lead to dystokia.

In conclusion here, the importance of correct feeding as a major determinant of profit, cannot be overstressed. Recent figures derived from the regularly conducted National Investigation on the Economics of Milk Production indicates that total food costs—grazing, conserved bulk foods and concentrates—account together for some 56% of all costs of producing milk. The significance of a figure of that magnitude—and therefore of this chapter—is clearly self-evident.

Chapter 3

WINTER FEEDING

THE LAST chapter dealt with the theory of feeding and how diets are calculated 'on paper'. Now we move to the most relevant topic for herdsmen, that of feeding in practice, starting first with the winter feeding period.

Before describing alternative systems in use for the feeding of forages and concentrates, it is necessary to outline several points of general stockmanship involved in the stockmanship of feeding.

- Allow time to observe the animals while they are feeding. Check that adequate trough space is available (recommendations given below), that minimum bullying is taking place, and, unless an ad-lib system is in use, that every animal comes to feed.
- Arrange regular feeding times (including weekends and when relief staff are on duty). Remember that cattle are creatures of habit and that the digestive system works more effectively if a regular pattern of feeding is maintained. As regular rumination is also required, rest periods between feeds need to be provided.
- Avoid over-feeding at one meal, especially with concentrates to high-yielding animals. The ideal 'meal' is first concentrates, then hay or a fibrous feed, followed by succulents.
- Allow the appetite of the animals to be satisfied. The theoretical diet may well provide essential nutrients but fail to fill the cattle, so that they remain restless and perform below expectation. Liberal straw bedding helps contentment as it allows hungry animals to eat some of the material before it becomes fouled.
- Check that the expected production level is being achieved. This involves measuring the production by recording the yield of milk or the weighing of growing stock. Recording of individual milk yield is ideal but much can be gained from using group yields (where applicable) or even daily herd yield. Adjustment of the diet will be required to correct variations from target. Note that

30

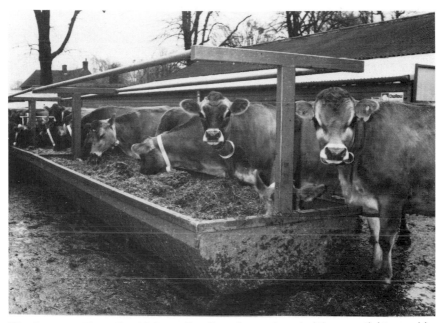

15. Jersey cattle eating high quality silage from a trough. It is essential to provide adequate feeding space if timid animals are to obtain satisfactory feed intake.

16. Cleaning out waste material from a concrete manger. It is necessary to ensure that diets are palatable and so, for example, when low-quality silage has to be offered, refusals needs to be removed regularly. Note the highly effective freeze-brand mark of 732.

17. A complete-diet feeder dispensing a mixed ration to a high-yielding group. It is important to drive accurately, so avoiding wastage—even on Sunday mornings! Note the piece of rock salt in the manger which the cattle lick to obtain sodium.

feeding levels also affect the quality of production, e.g. butterfat and solids not fat of milk.

- Ensure that diets are palatable. Food refusals may not be due to low appetite but to such problems as mouldy hay, dirty roots or butyric silage. Take trouble to prevent problem material, e.g. waste from the top of a silage clamp, being loaded into forage boxes or being allowed to drop down to a self-feed face. Clean out mangers regularly if low-quality material has to be offered. Small quantities of highly palatable materials such as molasses can be used to improve the palatability of a diet.

- Check the level of digestibility of the diet by the consistency of the dung, avoiding undue scouring or constipation. Buffer lush grass or kale with a little hay or clean straw.

- Watch out for laminitis and sore feet in high-yielding animals which are receiving a concentrate-rich diet. The addition of sodium bicarbonate to the diet (250–300 g per day) has been shown to be beneficial and it can also help a low butterfat problem.

- Avoid abrupt changes to the diet. The pattern of fermentation in the rumen is seriously affected especially if the concentrate:forage ratios are abruptly changed. Arrange for a change of diet to take place over several days.

- Minimise feed wastage, especially of expensive concentrates, but also of hay, silage or complete diet. Avoid over-filling low-capacity mangers, adjust the height of mangers where applicable and always drive accurately when dispensing from a forage box—even on Sunday mornings!

Planning the Feeding Programme

It is advisable to plan the feeding programme well in advance of the winter period, by first evaluating the available stocks of home-grown feeds. Any requirement for purchased feeds to supplement a short-fall can then be arranged in good time, hopefully avoiding the need to buy at the end of the period when prices are usually at the highest level.

Once the home-grown 'background' feeds are stored in a clamp, tower or stack, or as a standing fodder crop in the field, there is little the herdsman can do, at this stage, to influence feed quality. Ideally, he will have used his influence and helped out earlier in the season to obtain stocks of the optimum quality, by for example, rolling mole hills to avoid soil contamination of silage and ensured thorough sealing of the clamps.

The first part of the planning exercise is to quantify the amount of each type of feed available. With baled hay and straw, a count may need to be taken unless accurate records have been kept by the baler operator or by the staff involved in stacking. It is then necessary to weigh a representative sample of the bales, and by multiplication, to obtain a total weight. Few difficulties should arise weighing traditional small bales but it may be necessary to travel to a weigh-bridge to obtain the average weight of a few large bales.

Evaluating silage quantities is rather more difficult due to the fact that uneven shaped clamps are frequently formed. It is necessary to obtain the volume of material (m^3) by using a measuring tape (metres) and multiplying length × width × average weight.

The average weight of one cubic metre is 660 kg, although short-chopped material in deep clamps is heavier and lighter when un-chopped and in shallow clamps.

Feed required

The length of the average winter feeding period for budgeting purposes is usually 180 days although in northern regions it may well be over 200 days. Assuming that the home-grown background feeds are of sufficient quality to provide the nutrients for Maintenance plus 5 litres (M + 5) then the approximate requirements are:

Cow Live Wt. (kg)	Silage (kg)	Hay (kg)
600	7,650	2,550
400	6,000	2,000

More precise quantities will depend upon such factors as herd yield and attitude to concentrate feeding. High-yielding cows require less forage but animals offered only low levels of concentrate, but at the same time having easy access to high-quality silage, will eat as much as 8–9 tonnes each during the winter months.

If alternative foods are available a useful conversion is:

> *1 kg Hay* = 3 kg silage
> = 4 kg kale or beet tops
> = 5 kg mangolds
> = 2 kg straw
> = 1.5 kg chemically treated straw
> = 0.5 kg straw + 0.5 kg straw balancer
> = 1 kg brewers' grains + 0.25 kg cereals

Forage quality is best obtained from laboratory analysis, average value being given in Chapter 2, Table 2.5, although a useful guide to good material is:

> *Silage*—Greenish-yellow colour
> leafy, few flowering heads (cut early)
> sharp acidic smell
> not over-heated or with a butyric smell
> well made, absence of soil contamination
> difficult to squeeze out moisture by hand.
>
> *Hay*— absence of mould, a 'good' smell
> green or yellowish-green colour
> little loss of leaf
> no seed formed (cut early).

Once the amount and quality of the feeds available are known, then rations can be formulated for each class of stock, as explained in the last chapter. It is however, worthwhile having a computer calculation undertaken by a feed company or by an organisation such as ADAS or MMB. This is particularly useful if 'other' feeds such as beet pulp, brewers' grains or arable by-products are to be purchased. Given the prices of such feeds, the programme will calculate which one, or even a combination, is the 'best buy'.

FORAGE FEEDING SYSTEMS

1. Silage

There are a number of alternative methods of feeding silage from store, the suitability for a particular situation depending upon such factors as building layout, herd size, yield level and attitude to mechanisation of the management and dairy staff.

(a) Self-feeding is a satisfactory method of feeding silage, especially for medium-sized herds using minimal labour, providing that the layout is well planned and the slurry disposal system is satisfactory.

Short-chop material is ideal; it should not be stacked more than 2 metres deep unless some hand-cutting and forking down of the upper layers takes place. The optimum allowance of feeding face depends upon cow size, but primarily upon the time allowed for access. An adequate allowance for a small cow with 24-hour access is 20 cm but 80 cm will be required if large animals have to be batch fed at the face.

Herdsmen can control the face by using a good 'hot' electric fence or feed barrier to restrict the cow's access. The fence or barrier should be moved forward at regular intervals using knowledge of the performance of the stock and the availability of supplies. It is helpful to use marks along the walls of the clamp as an aid to monitoring use. When quality is low, it may pay to allow some rejection and remove the uneaten material daily for feeding to less critical animals. Problems can arise if two or more silage cuts have to be ensiled in the same clamp, particularly if the layout does not allow the first cut to be utilised when required for the freshly calved animals.

Heifers in their first lactation which are in the process of losing their calf teeth need to be watched carefully, especially if they have to pull out unchopped silage. Extra concentrate supplementation will be required for such animals until their broad teeth are fully cut.

Increased flexibility in cow management is possible with self-

18. Self-feeding of silage.

19. A front-end loader taking maize silage from the clamp. If secondary fermentation or rotting are to be avoided, this must be done with minimum disturbance to the remaining silage.

feeding if more than one feeding face can be provided, so allowing the herd to be subdivided. Such a facility enables the newly calved and first-lactation heifers to be managed in one group; it reduces stress on these critical animals, enabling higher peak yields to be obtained, as well as better conception rates.

Self-feeding is only suitable for high-yielding herds if facilities are available to feed concentrates other than in the milking parlour, i.e. in a manger or out-of-parlour feeders.

(b) Mechanical handling. With large herds, or on farms where the staff have considerable mechanical interests and skills, the handling of silage out of clamps by machine is a practical alternative. Many farms use a standard front-end tractor loader with a normal fork or bucket, others have a special grapple fitted to the loader, but some use specialised silage block cutters. With all types of machine the aim should be to remove the required amount of material with minimal disturbance to the remaining face. Secondary fermentation (or rotting) soon takes place when air is allowed into the face. The problem can be reduced by working methodically across the face of the clamp from one side to the other, by re-sheeting daily and by spraying the disturbed face with propionic acid from a hand-spraying unit.

All silage handling machines, but particularly block cutters, need to be regularly checked and serviced as reliability is an important feature of the system. A number of larger farms use industrial type loaders with the objective of increasing reliability as well as output.

When loading into a forage box or a complete diet feeder it is vital to avoid damaging the unloading and mixing mechanisms by first teasing-out the blocks and large fork-fulls of silage.

(c) Big bales. The development of big-bale silage is proving to be of considerable benefit to many beef enterprises but also to some dairy farms, and especially to smaller ones which have difficulty in justifying the capital cost of traditional silage making. High-quality material can be obtained by cutting and baling early, usually using a contractor, and by giving care and attention to sealing and stacking the bags, and to preventing vermin chewing the plastic sacks.

The availability of a quantity of bagged silage is proving, on many farms, to be a convenient method of buffer feeding cows during the summer and autumn months. Instead of opening a clamp to supplement inadequate grazing, large bales of silage can be fed from mangers located on concrete in the yards or even at pasture.

2. Hay

Hay is a most convenient form of conserved forage to feed, particularly in cowsheds but also in loose-housed situations where racks or mangers are available. Conventional sized bales are particularly convenient to handle and enable rationing to be easily undertaken. Large bales are increasing in popularity but need a specialised manger placed in a convenient position for loader filling, or a long feeding passage where bales can be easily unwrapped. Wastage can be a problem with hay feeding, especially when the quality and palatability are low.

3. Straw

Although straw is widely used in cattle enterprises for bedding and for feed, considerable developments are taking place in the methods of chemically treating the material to improve its digestibility. On-farm treatment of chopped straw with caustic soda is a specialised and potentially hazardous operation which is usually undertaken by contractors. Treatment with ammonia can be undertaken in specialised ovens on a batch system or by pumping the aqueous form into a stack of bales completely sealed with plastic sheets. Chopped or milled and treated material, as well as purchased cubed straw can be readily included in complete diets and bales used as a part or whole replacement for hay. The use of treated straw for beef cows and replacement heifers is expected to increase considerably in the future. It will no doubt also have an increasingly valuable role in dairy cow rations to optimise fibre levels and assist in the control of fermentation.

CONCENTRATE FEEDING

It was explained in the last chapter that the amount of concentrates to be fed to dairy cows depends upon the quality of the background feeds, the level of milk yield and the appetite of the animals which is affected by the stage of lactation.

A simple guide to rationing is shown in Table 3.1.

It is usual to feed all or part of the concentrate ration at milking time so that individual cows can be rationed according to such a scheme as in Table 3.1. Alternative systems of allocating concentrates, however, have been developed to avoid the need for frequent

Table 3.1   Rate of Concentrate Feeding

Milk from background feeds (kg per day)	Milk yield per day (kg)				
	10	15	20	25	30
	Concentrates per day (kg)				
M	4	6	8	10	12
M + 2.5	3	5	7	9	11
M + 5	2	4	6	8	10
M + 7.5	1	3	5	7	9
M + 10	—	2	4	6	8
M + 12.5	—	1	3	5	7

changes of the cows' ration and simplifying the job of the herdsman in frequently adjusting feeders at milking time.

Step Feeding

This is a system developed to simplify feeding but also to encourage high peak yields. Fixed daily amounts of concentrates are fed for predefined periods of the lactation, each usually lasting 8–12 weeks. The level adopted is related to the yield in early lactation (days 10–14) which has been shown to be a good guide to the anticipated potential of the cow, so long as she calved normally and has not been ill. If the herd is sufficiently large to justify subdivision into yield groups, allocation of background feeds and concentrates becomes relatively simple. Some facilities to enable the feeding of some concentrates out of the parlour will be necessary for high-yielding animals. It is usual to feed about 9 kg per day in the first two weeks

Table 3.2   Daily diets

Predicted peak yield:		35 kg/day			27.5 kg/day		
Predicted lactation:		7,000 kg			5,500 kg		
Forage quality	Period No.*	2	3	4	2	3	4
Very high	Forage DM (kg)	12.5	14.0	15.5	15.5	16.0	15.2
	Concentrate (kg)	7.5	4.0	2.0	1.0	0	0
Medium	Forage DM (kg)	7.0	7.5	8.0	8.0	8.5	8.5
	Concentrate (kg)	13.5	11.5	10.0	9.5	8.5	7.5

*Period 1 = 0–2 weeks; 2 = 3–12 weeks; 3 = 12–22 weeks; 4 = 22 weeks
Source: adapted from MAFF.

of lactation until potential yield is obtained. Table 3.2 gives examples of rations based on forage of varying quality.

Care should be taken to avoid reducing the plane of nutrition around the time of conception, particularly in a high-yielding animal which has had excessive liveweight loss in early lactation.

Flat-Rate Feeding

This is another system which avoids concentrate rationing of individual cows. It is a simple system in which a single level of concentrates is fed to the whole herd or to a group within the herd. High-quality forage (usually silage) is fed ad-lib so that higher-yielding cows due to their larger appetite can eat more silage.

The system is based on the fact that body weight and appetite vary through lactation and that reducing the level of starchy feeds in the diet increases the intake of high-quality forage.

Yield potential is again predicted (10–14 days) to enable an appropriate level of concentrate feed to be selected for the whole of the winter feeding period.

A guide to winter feed requirements for flat-rate feeding is given in Table 3.3.

Table 3.3   Winter feed requirements for flat-rate feeding

Breed	Calving season	Forecast yield (kg)	Daily concs. (kg)	Likely daily silage intake (kg)	Winter feed (tonnes) Conc.	Silage
Channel Islands	Sept/Oct	5,000	7	31	1.37	7.0
		4,000	5	31	0.98	7.0
	Jan/Feb	4,500	7.5	35	0.55	5.5
		3,500	5.5	33	0.40	5.3
British Friesian	Sept/Oct	6,500	8	38	1.56	8.6
		5,500	7	35	1.37	8.0
	Jan/Feb	6,000	8	46	0.58	7.4
		5,000	7	43	0.51	7.1
Holstein	Sept/Oct	7,000	9	39	1.76	8.8
		6,000	7.5	37	1.46	8.4
	Jan/Feb	6,250	9	46	0.66	7.5
		5,250	7.5	43	0.55	7.3

Source: Adapted from MAFF.

Complete Diet Feeding

This is a system of feeding which involves supplying cattle, on an ad-lib basis, with a uniform mix of chopped or coarsely milled forage, concentrates, cereals, minerals and other feeds. The mix is prepared and distributed using a mixer wagon which is normally fitted with load cells so that accurate weighing of the ingredients can take place. The herd, which has to be sufficiently large to justify the capital cost of the equipment, is subdivided into yield groups with diets prepared with differing energy densities according to the requirements of the average animals in each group. A small proportion of the herd may become over-fat but are valuable culls.

Table 3.4   Ration formulation

Stage of lactation	Approx. DMI (kg/day)	Suggested M/D (MJ/kg DM)	Crude protein %	Forage DM % total DM
Early	19.0	11.5	16	35–45
Mid	16.5	10.8	14	55–65
Late	11.0	9.8	12	70–80
Dry	11.0	9.0	12	80–90

Many farmers using C.D.F. have discontinued the feeding of concentrates in the parlour although others feed a small quantity in order to encourage reluctant cows to enter for milking. The system is capital intensive, and can be efficient in labour use although the 'feeder' does require a high level of skills in stockmanship, mechanisation and even mathematics! It is a system which offers considerable scope to increase milk yields and compositional quality but to be economic, needs very careful monitoring.

Out-of-Parlour Feeders

This is another capital-intensive system of feeding concentrates but one which is being increasingly utilised in medium-sized high-yielding herds where C.D.F. is not feasible.

It is a system which fits well with the self-feeding of silage, especially in building layouts where it is inconvenient to locate mangers for concentrate feeding. The system involves a number of feeding stations (each serving 25–30 cows), which are best sited between the forage feeding area and the cubicles or bedded yards.

20. An out-of-parlour feeding
unit with three stalls and
integral bulk-storage hopper.

21. Cows using the out-of-parlour feeders. The spring-loaded rear gate prevents
other cows entering the stall while an animal is using the feeder.

Cows are fitted with transponders on a neck collar and after a period of training soon get used to entering the stalls of the feeder and receiving their ration of concentrates. The current type of system being installed rations the cows individually, using a microcomputer which also displays the cows' daily consumption as well as the total daily usage. The system can be of help to herdsmen by displaying the number of an animal which has not eaten her allocation because such an animal may well be 'off-colour' or in oestrus.

Manger Feeding

Concentrate feeds over and above that received in the parlour at milking times can be fed to high-yielding cows in an outside manger or at a feed fence. Bagged food is a convenient method of handling and certainly of rationing, although various types of trolley and mechanical dispenser can be used. A facility to be able to yoke the animal is ideal; it is a high capital cost but prevents bullying and can also be useful to herdsmen while undertaking many of their husbandry tasks, and even for AI. When feeding at a manger without yokes, the ideal feeding space for friesian type cattle is 65–70 cm.

22. A good feeding manger with self-locking yokes ideal for individual rationing of concentrates or carrying out husbandry tasks such as condition scoring.

This amount of space will require the cattle to pack together so that they are less likely to back-out knowing that they will have difficulty finding an alternative place along the line. The provision of a manger feed late at night is a good method of increasing nutrient intake but disturbance to the herd at this time makes the task of oestrus detection much more difficult.

Chapter 4

SUMMER FEEDING

HERDSMEN, WITH few exceptions, look forward to the spring months when the cattle can be turned out to pasture and the work load associated with winter chores is reduced. As grazed grass forms the major part of the diet of beef and dairy cattle during the summer months, it is important for herdsmen to be aware of the factors which affect grass growth and its utilisation by the stock.

Following the relatively precise feeding systems of the winter comes the difficulty of dealing with a material which is highly variable in terms of both quality and its availability. The reasons for this variability are dealt with in this chapter, together with the practices which can be undertaken to get the most out of what is the cheapest type of feed. Before and during the season many important decisions have to be taken such as which fields to graze, cut or fertilise. In many of these situations the decision will have to be taken by the herdsman alone, but in others, it will be the management, hopefully in consultation with the herdsman, who decides on a particular course of action.

Factors Determining Grass Growth

Before dealing with the alternative systems of grazing and the factors which influence efficient utilisation, it will be helpful to consider a number of factors which determine grass growth.

- *Farm Location.* Although temperature and radiation from the sun influence growth, the most important factor in Britain is the rainfall during the growing season (April to September). Soil type is also a key factor, especially its water-holding capacity. The two factors of rainfall and soil type can be used to predict the grass-growing potential of a particular farm or field. A system of classification has been worked out at the Grassland Research Institute based on site classes from 1 (very good) to 5 (poor). By reference to Table 4.1 the grass growing ability of a particular location can be obtained.

Table 4.1 Grass growing site classes

| | Average rainfall (April–Sept) | | |
| | More than 400 mm (16 in) | 300–400 mm (12–16 in) | Less than 300 mm (12 in) |
Soil texture			
Clay loams and heavy soils	1	2	3
Loams, medium-textured soils; deep soils over chalk	2	3	4
Shallow soils over chalk or rock; gravelly and coarse sandy soils	3	4	5

The total dry matter yield from each site class from three cuts and a subsequent grazing were:

Site class	DM yield (t/ha)
1	12.7
2	11.6
3	10.5
4	9.5
5	8.4

It is interesting to note in Figure 5 that most of the difference in yield occurs after June when poor sites suffer most from mid-summer moisture stress.

- *Types of Grass.* The species of grass being grown does not have such an important effect on yield as many farmers and stockmen think. Ryegrass, timothy, meadow fescue and clover are certainly preferred species but good yields of digestible forage can be obtained from permanent swards which have a high proportion of useful indigenous grasses. Over half the lowland grass used by dairy cows in the United Kingdom is unsown or is more than ten years old. Apart from the swards with a high weed population it is usually worthwhile improving and maintaining pasture quality by management rather than incurring the high cost of reseeding.

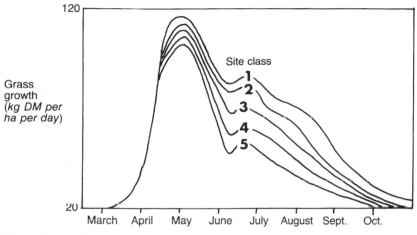

Source: Thomas and Young.

Figure 5. Seasonal pattern of yield of S23 perennial ryegrass.

Sown grasses do provide earlier spring growth (especially Italian ryegrass), heavier silage cuts and offer less risk of parasites to grazing stock. Permanent pasture, however, is more drought resistant, withstands treading much better, especially in high rainfall areas, and is well suited to small intensive farms where the risk of a failed ley can be a serious matter.

Clover has higher intake characteristics than grass and saves on nitrogen fertiliser costs. It does however contribute little to output in early season or on farms when more than 200 kg/ha N is used.

- *Lime Status.* The pH of the soil should be monitored by regular tests and lime applied to maintain a level of 6.0. The better species tend to die out if the soil is allowed to become too acidic and the response to fertiliser is reduced. The more nitrogen is used the more frequently lime will be required.

- *Weed Control.* Docks, thistles and buttercups are common problems in permanent grass as are arable weeds in many newly established leys. M.C.P.A. and 2.4-D are suitable chemicals to use for control where no clover is present, otherwise M.C.P.B. and 2.4-DB should be used. (Docks may need several treatments.)

- *Phosphate and Potash.* Grazing allows much of the P and K to be returned to the soil so that on fields which have been intensively stocked, annual dressings of 30 kg/ha P_2O_5 and 40 kg/ha K_2O are usually adequate. When fields are frequently cut for conservation increased K is needed at a level of 0.5 kg K_2O for each kilogram of

N applied. Soil analysis should be undertaken to monitor P and K levels. If slurry is applied the above levels can be reduced or even eliminated. Large single dressings of K should be avoided otherwise 'luxury' uptake occurs which is not only wasteful but also increases the risk of hypomagnesaemia.

- *Nitrogen.* Because this is the major nutrient affecting grass growth its usage has a marked effect upon the productivity of a cattle enterprise. Although the average application on UK dairy farms is little over 150 kg/ha, the economic optimum or target levels vary between 300 and 450 kg/ha depending upon site class. See Table 4.2.

Table 4.2   Target levels of nitrogen use

Site class	kg/ha
1	450
2	400
3	350
4	325
5	300

The distribution of yield through the season can be influenced by the timing of applications. Heavier mid-season applications will improve production at this time so long as adequate soil moisture is available. The first application of the season should ideally take place before growth commences, with subsequent dressings to grazed areas every 3–4 weeks. The level applied should be 1.7 to 2.5 kg/ha per day depending upon site class. For silage cuts 120–150 kg/ha is recommended for the first cut and 100 kg/ha for subsequent cuts.

- *End of Season Management.* There is a tendency for grass, especially in sown leys, to continue growing well into the autumn when the economic utilisation by dairy cows becomes infeasible. Leaving this uneaten material in situ can cause problems, especially in a hard winter when a high proportion of the plants can be killed out. The ideal management tool is a flock of grazing sheep— although not all dairy farms have access to such a facility—or appropriate fencing. Grazing with young stock may be a possible solution but to maintain adequate growth rates at this time supplementary feeding will almost certainly be required. Whatever type of animal is used to remove surplus grass they should be taken off the area, ideally before Christmas and sooner if very wet conditions occur.

Forage Quality

Having grown the grass, then comes the challenging task of utilising it efficiently for grazing. The first factor is that quality varies markedly throughout the season. This is measured in terms of digestibility of the material (D) which is the percentage digested by the animal. Young very leafy material in early spring can be as high as 80 D so that supplementation with a fibrous feed such as hay or straw is essential otherwise it moves through the gut too quickly and inefficient digestion takes place. The aim should be to present the animal with material that has a good proportion of leaf to stem and rarely falls below 70 D. The D value as well as level of crude protein falls as the grass begins to flower, with a mature crop having a digestibility of below 60 D. Each variety has a characteristic relationship between digestibility and stage of growth, e.g. Timothy declines well before ear emergence. Autumn grass is not digested as efficiently as spring grass due to the lower sugar levels (young spring stems have a higher sugar content than leaf).

Regular topping of pastures removes any seed heads and ensures that higher quality material is available at the next grazing. This is a suitable job for herdsmen to undertake; it not only gets them out of the buildings into the 'fresh air' but also allows them to tidy up the grazing which can be good for morale. If the cows are in the next field or paddock it is often possible, whilst driving up and down, to observe their behaviour over an extended period.

Grass which becomes soiled by urine or dung is unpalatable for several weeks and tends not to be eaten. This problem appears to be more serious when stocking rates are at an average level rather than when low or very high (presumably the cows get used to soiled grass). Unless heavy rain soon follows, harrowing of dung pats can make the situation worse.

Amount of Grass on Offer

Thick dense swards of reasonable length allow the grazing animals to consume a full diet with minimum movement and exertion. Sparse over-grazed crops however require the cattle to work harder and after several hours they become fatigued, discontinue grazing, so that appetite is not met. Production is then seriously affected as maintenance requirement is increased at a time when nutritional intake is low. Grass consumption on a good sward can be as high as 70 kg per day but is only half that amount on a thin poor sward. Growth of grass can be erratic due to rapid weather changes, so it is an important job for the herdsman to check on the amount of grass available. On wet and windy days the animals will spend time sheltering to the detriment of grazing, and when they do graze, dry matter levels will be low. A useful task is to measure the length of grass after grazing with a ruler or perhaps using marks on the wellington boot! Ideally as the cattle leave a rotationally grazed area the average height of remaining grass should be between 8 and 10 cm (about 4 in). In set-stocked areas, production will be reduced if the average height of grass falls below 6–8 cm (3 in).

Stocking Rate

This is another key factor affecting the profitability of grassland but one which can be agreed and set before the grazing season begins. Low stocking rates lead to uneaten grass which dies and is mainly lost. Remaining material becomes stemmy, digestibility falls so that intake and milk yields also fall. Too high a stocking rate leads to overgrazing and also to the lowering of intake and performance.

Grazing stocking rates need to fall during the season to match the natural decline in growth and quality of grass. On a class 1 site six cows per hectare can be carried up to mid-May but only 2.5 in the late season. Areas which have been earlier cut can be added to the grazing area as the season progresses. Conservation is not only a means of producing stocks of winter feed but is a vital tool in the management of summer grazing.

Irrigation is a technique which a few farmers are able to use to improve stocking rates or, in other words, reduce the risk of a high stocking rate in dry years. It is in many ways a method of improving the site classification but is usually obtained at high capital cost and on some farms, which use hand-moved pipes, has a high labour requirement.

Shortage of summer fodder is a frequent problem for herdsmen on poor grass growing farms, but one facing most at some times. It is dealt with in some situations by the purchase of feeds such as brewers' grains or arable by-products. Others buffer the grass with silage (big bales being most convenient), hay or straw. Increased concentrate feeding is frequently involved but this subject will be discussed below. Lucerne with its deep rooting system can be useful under such conditions as it continues to grow during dry periods but it can only be reliably grown on free draining soils of high pH in the southern and eastern counties. Rape and early planted kale are other alternative feeds which can supplement grass in the summer and early autumn months. Germination and growth of these crops is also highly weather sensitive so that failures are not unknown.

The Milk Potential from Grass

It has been explained that when grass is available of the appropriate quality and quantity the cow will satisfy her appetite and produce to her potential. The situation can arise in which the cows are yielding less than the amount that the grass is capable of supporting. Unless

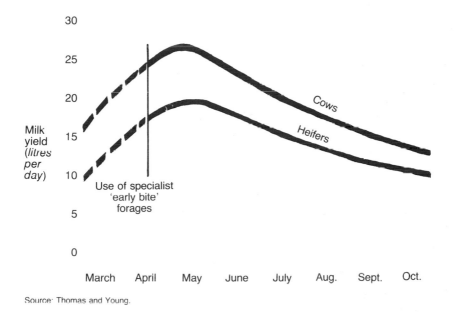

Source: Thomas and Young.

Figure 6. Potential daily milk production for ad libitum high-quality grazing.

stocking rates are increased accordingly, grass in these situations is wasted. In herds which have a wide spread of calving dates, the difficulty arises in balancing the needs of individual animals with stocking rate and with fertiliser use. Subdivision of a herd into yield groups does help this situation but at the same time it increases the time spent by stockmen in bringing cows to and from pasture. The potential milk production from high-quality grazing is shown in Figure 6.

Supplementary Concentrate Feeding

The feeding of concentrates to dairy cattle during the summer months, although practised by many producers, is seldom a profitable activity. It may be justified for short periods when grass growth is slow, but when adequate forage is available the animal substitutes concentrates for grass and produces very little extra milk. When grass is in short supply body condition can soon be lost which has a long-term effect on cow performance, so that this is another situation in which concentrate feeding is justified.

High-yielding spring calvers will also justify concentrate feeding if their yield level is above the potential of the grass. This will be more so with fresh calvers whose appetite is still at a low level. There is no fixed rule for the rate of feeding concentrates in the summer but the type of concentrate to be fed is important. The protein in grass is highly degradable, particularly in mid and late season so that high-yielders require undegradable protein supplementation. Energy is usually limiting for high-yielders and also in order to maintain adequate butter fat levels the type and content of the oil in the diet have to be taken into consideration. In terms of minerals, grass is usually lacking in phosphorus, and in some cases in trace elements, and of course magnesium.

The general conclusion about the feeding of concentrates at grass is that the herdsman should have confidence in the grass, give it a chance to produce the milk and only supplement with concentrates at critical times. The rate of decline of milk yield, as well as the body condition of the animal, should be regularly monitored and buffer feeding introduced, ideally from forage type feeds, if there is any sign that grass is not fully meeting requirements. It may even be a sensible precaution to have some buffer feed on offer, i.e. a rack of hay or straw, throughout the season at some convenient place in the dispersal area.

GRAZING SYSTEMS

Several methods of controlled grazing are in use for dairy cattle, all of which aim to provide the animals with adequate forage but at the same time minimise wastage.

1. Strip Grazing

This system involves the use of an electrified fence which is moved once or twice daily to provide the herd with a clean strip of forage. It

23. Cows grazing a good leafy sward.

24. Strip grazing behind an electrified fence.

is an efficient system of rationing and generally avoids wastage, although a back fence is necessary in a large field to prevent eating of the re-growth. The system is rather labour consuming for herdsmen but the biggest problem is usually the one of deciding how far forward to move the fence. Providing too much grass causes selective grazing and wastage but too little fails to meet appetite and the cows are also more likely to break through the fence! Poaching can be a problem in wet conditions on heavy soils which is more likely to occur when the system is used to ration kale or roots. Surplus grass can be taken more easily for conservation than with other systems and the provision of water troughs seldom causes a problem.

There is some skill involved in putting up an electric fence efficiently. Ensure that the end posts are firmly in place then have the wire as straight and tight as possible. The height of the wire is usually set by the use of 'standard' pig-tail posts and it is normally 80–90 cm from the ground. Mains supply fences should be installed and maintained according to the manufacturers' instructions and batteries replaced in portable units, as required.

2. Paddock Grazing

With this system, permanent and semi-permanent fences are used to subdivide the grazing area into a number of paddocks, so reducing the labour requirement for moving fences during the summer. The system is based upon graze, fertilise and rest for a period until grazed again. The interval between grazings may be as short as twenty days in the early part of the season, extended to over forty days in the autumn. At the peak of growth, surplus paddocks can be cut but problems do arise with large machinery operating in small and awkward shaped paddocks. When laying out the fences it is worthwhile arranging for the paddocks to have parallel sides which make fertiliser application, harrowing, topping and irrigation much more efficient. Ideally, the width of each paddock should be multiples of twice the working width of the fertiliser applicator (to avoid empty runs back to the gate).

Some farms operate with many paddocks, whereas others have a few larger ones and some even strip-graze within the paddocks. Daily milk yield can vary considerably when using a few large paddocks, this being an example where access to a buffer feed minimises such problems. Adequate water should always be available, perhaps one trough in the fence line shared between two paddocks. Good roadways are advantageous, not only for vehicular traffic, but also for the efficient movement of the herd. The system is tending to lose popularity due primarily to the high cost of fencing and water supplies.

3. Set-Stocking

This is the traditional system of grazing which allows the cattle free access over a set area of pasture. It has been more recently developed to carry stocking rates almost as high as in rotational grazing. Nitrogen is applied in blocks or over all the area, usually on a 3–4-week cycle. The size of the area can be adjusted after each silage cut and another form of stock, e.g. sheep and lambs, drafted in for a short period if undergrazing is a problem. In order that the grass does not initially 'get away' from the cattle, turn-out has to be early with buffering by winter rations taking place until grass growth is balanced with herd need.

One of the major limitations to this system for herdsmen is the time involved collecting the cows, especially with the larger herds in big fields. In some situations the problem is eased by operating with two fields, one for day use and the second more convenient one for night grazing.

Milk production on each of the three systems is very similar, so that the one selected for a particular herd will depend upon farm layout, but usually more upon the preference of the people involved. There is perhaps one exception in that rotational systems do have a slight edge over set-stocking for high-yielding spring calvers.

Water Supplies at Pasture

Ideally every field or paddock should have a water trough and it should be sufficiently large so that at least 10 per cent of the group can drink at any one time. The average consumption varies according to the weather conditions and milk yield, but on a hot summer's day high-yielders can drink as much as 180 litres each. On farms where herd size has increased dramatically it will usually be necessary to increase the diameter of the supply pipe from the customary ½ inch (12.7 mm) to ¾ inch (19 mm) or even 1 inch (25.4 mm).

Zero-grazing

A small number of milk producers with unhandy farm layout or perhaps where a motorway has subdivided the farm, cut and cart fresh grass to the herd throughout the summer months. Improved

25. Collecting cows from paddock grazing. This is an easier task for the herdsman than collecting from a set-stocked area.

26. A large-capacity water trough, allowing at least 10 per cent of the herd to drink at any one time.

levels of utilisation occur due to the elimination of selection and fouling, so that higher stocking rates are achieved. There is no need for fencing or water supply but the available mechanical skills do need to be considerable. It is a system which can create considerable stress on cows and machines, as well as on herdsmen and management, particularly on farms where the slurry system has difficulty in coping.

Storage Feeding

Using conserved feeds as the basis of feeding throughout the year allows for high stocking rates and avoids the daily chore of cutting and carting grass. It is a system widely practised in such countries as the USA and Israel where regular supplies of high-quality feeds (e.g. lucerne hay) can be purchased economically. It is now being used in the UK by only a few producers but many more are showing interest in storage feeding due to the increased control which is possible over feeding and milk production. Cow health is an obvious potential problem as is machinery reliability, but it will no doubt be a system more widely used in the future.

A compromise situation could have more attraction to herdsmen, i.e. storage feeding by night and grazing by day. The cows are then handy for morning milking, which is especially valuable when they are milked three times a day. Temporary grass shortages are easily compensated by the animals increasing their silage (or hay) intake at night. Increased silage quantities are obviously involved but it is a system offering considerable flexibility in that it can be readily introduced or discontinued according to silage/grass availability.

Folding Green Crops

The strip grazing of kale, roots or even sugar beet tops is a method of extending the grazing season. With kale it is preferable to provide only two or three rows rather than a deeper strip, so as to minimise wastage. An alternative system is to cut and cart the kale, ideally using an offset, side-mounted forager to minimise soil contamination.

When grazing green crops it is also preferable to provide a grass or stubble lie-back area and to avoid long walks to and from the field. There is some risk with digestive upsets in frosty weather so that a hay feed should be given before they are allowed to graze kale.

Forage Utilisation

It has been shown that the main aim in summer feeding is to make the full use of pasture in conjunction with other supplementary feeds. A simple estimate of the energy contribution of grass and forage on a farm is contained in the concept of Utilised Metabolisable Energy (UME). This figure is obtained by subtracting from the total energy required by the animal, the energy supplied by the concentrates and other purchased feeds. The UME of a farm with well-managed grassland is 75–80 GJ per hectare (1 GJ is equal to 1,000 MJ).

Chapter 5

COW MANAGEMENT

HAVING DEALT with the important topic of feeding in the last three chapters we now move to an equally important one for hersdmen— that of managing the cows through the production cycle. Management is a word used in several contexts but what is meant in this situation is the provision of the best possible working conditions for the cattle. This is the day-to-day tactical or operational management rather than the business or financial management in which the herd owner or manager is more likely to be concerned.

Herd management, as with any management task, involves planning, making decisions and control, but it also involves herdsmen taking action at the right time, without unnecessary cost, to ensure that the cows are healthy, well fed and producing at the optimum level. It will usually involve using the resources that are available such as the buildings, feed supplies, other staff and machinery, but it may from time to time require bringing in extra resources such as the vet or a freeze-branding contractor. Many of the tasks will ideally have been planned well ahead and agreed with the owner, others, however, will be unexpected (a cow calving three weeks early) or a real emergency (several bloated cows). Herdsmen need to know how to cope in all such situations and this will come with experience helped (hopefully) by reading these pages!

The varying needs of the dairy cow through the production cycle will be considered in this chapter but first it will perhaps be helpful to cover several techniques involved in efficient herd management, i.e. identification, herd grouping and body condition scoring.

Identification

All cattle, but especially dairy cows and youngstock, need to be clearly marked for easy and accurate identification. For health control reasons there is a statutory requirement for all cattle to be

27. A metal ear-tag being placed in the right ear of a calf, as required before fourteen days of age.

28. A plastic ear tag carrying the number of the animal (42), the initials of the sire (R.G.) and the year of birth (1973).

permanently identifiable with an approved type of metal tag or ear tattoo. Each herd has a mark allocated by the Divisional Veterinary Officer in a letter–number combination on a county basis. The ear-tag has to be placed in the right ear of the calf before it is fourteen days of age.

Large plastic ear-tags can be a most satisfactory method of identifying calves and youngstock until they are of a suitable size to be permanently marked, for example by freeze branding. Even with adult cattle, a large ear-tag can be helpful when observing the animals from in front, as at a feed fence. Interest for herdsmen in the subject of herd breeding is often enhanced by adding to the tag the initials of the animal's sire as well as the year of birth. Use of different coloured tags for each breeding season or progeny group are alternative methods of identifying 'similar' animals.

Freeze branding is now the most common method of identifying cattle, especially for black and white cattle, this being preferred to plastic collars, neck chains or ankle straps. The system of milking determines the site of the marks, but for herringbone milking the rear of the flank is usual (but on a black patch). Freeze branding is a patented technique so it should only be carried out by the licensees of the patent holder. In this form of branding the pigment-forming cells of the hair follicles are destroyed by the low temperature of the branding iron. The subsequent growth of hair is white and hopefully this persists for the lifetime of the animal. Increasingly, branding is being carried out at the yearling stage so that the numbers are ready for use during the first service period.

The trend towards electronics in dairying involves the use of automatic identification with transponders attached to a neck collar. The energising and interrogation equipment is located at the stall in the milking parlour, at the out-of-parlour feeding stations or at the walk-through weighbridge. One can imagine that before too many years it will be commonplace for animals to have a permanent life-time transponder 'implanted' into the rumen. Such a device should be a deterrent to cattle rustling and be of value to a national data collection scheme if and when one is established.

Temporary marking to identify the cows which are undergoing antibiotic treatment or requiring drying-off can be achieved by using coloured grease crayons or aerosol sprays. Adhesive tape can also be used on the tail for differential feeding or other management purposes, for instance a cow to be culled, so as not to be re-bred. The major problem with such tail tapes is that if they are fitted too tightly the blood supply to the switch is affected with serious results.

29. A freeze-brand mark ideal for general identification but especially at milking time.

30. A transponder fitted to a neck collar for automatic identification.

31. Ankle strap.

32. Neck chain.

Grouping of Cows

Reference has already been made in earlier chapters to the benefit of subdividing a herd into groups for differential feeding and management purposes. With cowshed housing the ultimate in individual feeding is possible as it is with yokes or individually programmed out-of-parlour feeders. In loose-housing situations grouping has much to offer, especially in herds over 80–90 cows and when rationing of concentrates in the parlour has been replaced by a block-feeding scheme such as complete diet feeding or step feeding.

Grouping allows one or more dry groups to be formed, perhaps one of first lactation heifers as well as subdivisions of the main milking herd. The number of groups formed in a specific situation will depend upon such factors as herd size, calving spread, building layout and, especially, the feeding arrangements. A 150 cow herd with a fairly tight calving pattern could have two groups of 75; a herd of 300 could have five groups of 60. As explained in Chapter 1, less bullying appears to occur if group size is kept below 80 but the optimum number should be in multiples of parlour capacity (or of the sides of a herringbone), e.g. 64 or 72 for an eight-stall-sided parlour.

33. Subdivision of a herd into groups for feeding and management purposes.

The most debatable point is whether to group by milk yield or date of calving. In the latter situation cows stay in their group, whereas in the former it is necessary from time to time to move animals between groups. Opinion is equally divided as to whether cows lose milk when moved, but in most situations this is to be expected as the level of feeding is lower for the animals moved 'down'. The author has found few problems in moving so long as several cows are moved together and if moving is arranged at a time of low herd activity (late afternoon) rather than at times of high competitive activity (morning feed time). In cubicle houses it is a considerable advantage if there is a passageway from feed to rest areas at each of the buildings. In this situation newly introduced cows can keep on the move and not be trapped in a 'dead' corner, as is frequently the case if only one passageway is available.

Some herdsmen (and owners) consider the extra time involved in moving groups to and from the parlour, especially in the summer months, to be unjustified. Before readily dismissing the technique they are strongly recommended to give it a try and so evaluate the benefits, especially those of reduced stress on the cow. It may well take those few extra minutes to change groups, but this time out of the pit is well spent, having a short break from the milking routine. Morale can be lifted considerably by, for example, washing down the standings after one group and starting the next group—which has a different level of production. Milking need not be a job just to be rushed, if better oestrus detection is achieved with smaller groups, the extra time involved in changing groups may be very well spent.

Body Condition Scoring

Reference has already been made in earlier chapters to the need for herdsmen to monitor the body condition of their cows during the year. The technique of body condition scoring has been developed as a standardised method of assessing the body reserves of the live animal. It is a very useful technique for herdsmen to use in controlling feeding and management of the herd so that it is one with which they should become familiar. This section is written to help the understanding of what is involved, but some training and regular supervised practice will also be required before a herdsman could become proficient.

As it is the level of fat cover over the muscles that is being assessed it is necessary to be able to distinguish between the firm feel of muscles and the softer feel of fatty tissue. The ideal sites for assess-

34. Body condition scoring. The level of fat cover around the tailhead is being assessed.

ment are the loin area between the last rib and the hip bone, as well as on each side of the tailhead. With animals in poor condition the loin should be used but with fatter animals a more accurate assessment can be obtained from the tailhead.

In order to undertake the task it is necessary to have access to the side and rear of the relaxed animal so that the ideal location for the job is in the cowshed, crush or AI stall. On the loin the degree

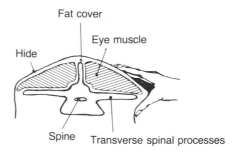

Figure 7. Assessing condition using the loin region.

of prominence of the spinal processes is obtained by feeling along the backbone but also by placing the thumb on the transverse process as shown in Figure 7.

Around the tailhead, the level of fat cover is assessed with the fingers and a score awarded from 1–5 as indicated below, to the nearest half point.

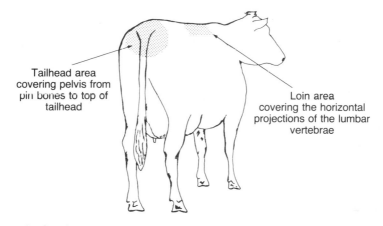

Tailhead area covering pelvis from pin bones to top of tailhead

Loin area covering the horizontal projections of the lumbar vertebrae

Figure 8. Scoring areas.

It is necessary, if there is a difference in scores between the tailhead and loin areas, to adjust the tailhead score accordingly, but by no more than half a point.

For example:

Tailhead score	Loin score	Difference	Adjustment	Adjusted tailhead score
4	2½	1½	− ½	3½
1½	2½	1	+ ½	2
3	2½	½	0	3

Condition Score
1. The individual spinous processes are sharp to the touch and easily distinguished, there is no fat around the tailhead.
2. The spinous processes can be identified individually but feel rounded rather than sharp. There is some cover around the tailhead.
3. The spinous processes can only be felt with firm pressure and the areas on each side of the tailhead have some fat cover.

4. Fat cover around the tailhead is easily seen as slight mounds and is soft to the touch. The spinous processes cannot be felt, even with firm pressure.
5. Very rare in dairy cattle, the tailhead and hip bone are almost covered by fat.

Minimum acceptable condition scores for each month of the productive year of the dairy cow are indicated in Figure 9.

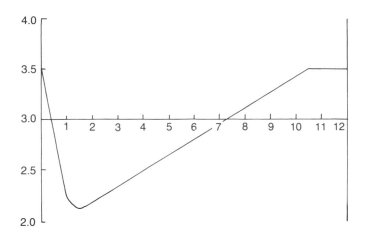

Figure 9. Minimum acceptable condition scores.

Cow Management through the Production Cycle

Late Lactation

At first thought, it would seem more appropriate to begin a description of the management of a cow through the production cycle by dealing with calving, or perhaps the dry period. However, studies of the results of nutritional research and talks with the owners and stockmen of many 'top' herds, have made it clear that it is in late lactation when the most effective influence on the production cycle can be undertaken. It is for this reason that management in mid lactation will be considered first.

During the second half of lactation, cows should be restored to their ideal body condition (3½ by drying off). This will require, for a mature Friesian type cow, a gain of 100 kg comprising: 50 kg for the new calf and uterine fluids and 50 kg to restore the weight loss of early lactation which in 150 days is 0.66 kg/day. For a young cow which is still growing, a further 50 kg will need to be put on in the last 150 days of lactation, increasing the gain to 1 kg/day. Research work has indicated that it is much more efficient in the use of nutrients to obtain this liveweight gain whilst the cow is milking, rather than when she is dried off. Once the ideal condition is reached it should be maintained by adjusting the feeding rate. Excess body condition is associated with fatty liver problems and is detrimental to milk production, so that it is important not to over-feed even in late lactation. Hence the importance of effective body condition scoring.

Dry Cow Management

The dry period should extend for 40–70 days to allow the udder tissue time to regenerate before the demands of the next lactation. It is preferable to dry off cows abruptly at the end of lactation, treating all quarters with long-acting antibiotics. This treatment controls persisting mastitis infections and prevents most of the new infections that would otherwise occur in the dry period. At the last milking the udder should be milked out fully, the teats thoroughly cleaned with white spirit and the antibiotic infused. The teats should then be dipped in iodophor solution and the udder left to stock up. Feeding for the next few days should aim at minimising milk production by arranging for grazing a bare field or, if yarded, providing straw or poor-quality hay. Cows should be watched closely until the udders contain no milk. Any quarter looking swollen should be checked for hardness and, if there is any, it should be milked out and reinfused with antibiotics. Fly control is important during the summer months so that bi-weekly spraying, or the fitting of repellent ear-tags, should take place.

Managing dry cows which are in good condition after milk secretion ceases is not a difficult task. The energy concentration of the diet should be kept low—i.e. there should be a high forage intake—in order to keep the rumen extended. It is considered that this is the ideal preparation for a good appetite after calving when an energy-rich diet can then provide maximum nutrients.

Exercise is good for heavily pregnant cows in, for instance, reducing the incidence of swollen and inflamed udders. Grazing is ideal, or if the cows are housed, it is preferable to use a reasonable sized yard where some regular movement can take place. When individual calving boxes are in use it is preferable not to move the cow into this facility until very near the calving.

A few cows and heifers may develop swollen and inflamed udders well in advance of calving if 'steaming up' with concentrates takes place at too high a level. This congestion can be relieved by reducing the energy density of the diet but it may also be necessary to ease some milk from the hard teats twice a day until twelve hours before calving. This milk, which is true colostrum, will be lost as a vital feed for the calf unless it is stored, preferably in a refrigerator.

Calving

The birth of the calf is a critical time for a cow so that herdsmen need to understand what is involved in the process if they are to provide appropriate care and attention. Although calving is a natural process which ideally takes place unassisted, close observation is required in case difficulties arise. Heifers calving for the first time tend to have more problems than older cows or suckler beef cows which, unless in-calf to a 'large' breed, usually calve easily.

Animals which do calve naturally and expel the after-birth satisfactorily, settle better into lactation and re-breed with fewer problems. There is a big temptation for keen but inexperienced herdsmen to interfere too much with calvings, and in most cases to start pulling the calf too soon. This premature interference can easily cause tissue to be torn or bruised which leads to genital infection and subsequent difficulty in re-breeding. Patience is therefore the number one factor in calving. There are, however, numerous other factors involved in efficient management of calving which will now be discussed.

It is quite satisfactory, in reasonable weather conditions and especially in daylight hours, to allow the calf to be born at pasture. It can, however, be a time-consuming and even frustrating task for a single-handed herdsman to move cows and newly-born calves from field to buildings. Housing in a calving yard or box is more convenient, especially if a complication does arise and the cow has to be restrained for examination and further assistance. It is unwise to calve a cow tied by the neck with a chain. If the cow has to be tied, a halter is preferable, tied with a knot which can easily be released.

The signs of calving approaching are:

- increasing distension of the udder
- stiffening of the teats
- in some cows the 'running' of the milk.

The animal tends to detach herself from the remainder of the group and when in the field may select a far from ideal site against a fence or ditch. The best guide to time of calving is to feel for a

35. Checking for the imminence of calving. This is indicated by a loosening of the muscles around the tail and the appearance of sunken areas between the tail root and each pin bone. It usually takes place 12–24 hours before the birth.

loosening of the muscles around the tail and for sunken areas between the tail root and each pin bone, which normally takes place 12–24 hours before the birth.

Normal Calving

When calving is more imminent the cow is uneasy and often in some pain. She tends to stay on her feet but continues to eat and drink. At this time the uterus is contracting, the water bag pushes through the cervix and the strains get more pronounced. This stage can take two to three hours with a cow and as long as six hours with a heifer.

More vigorous straining indicates the start of the next stage of the process. The calf twists a quarter turn, the head comes through the cervix, the water bag may burst and the feet can usually then be seen. This stage can take up to four hours, even longer with a heifer, which tends to tax the patience of an inexperienced herdsman. As the chest comes through the cervix, mucus is squeezed from the mouth and nostrils which clears the respiratory passages ready for the calf to breathe.

36. Calving.

37. Satisfactory delivery without assistance.

Once the calf is born the cow normally gets to her feet and begins to lick the calf vigorously. The calf staggers to its feet and soon begins to suckle—which helps the afterbirth to be released.

Close observation is required during the birth but so long as some progress (however slow) is being made, there is no need to interfere. The foetus continues to receive nutrients and oxygen via the navel cord until this is ruptured in the final stages of calving. Pulling the calf too soon is the main cause of torn tissue, as the cervix may not then be fully dilated. It is the intermittent pressure of the calf's head which stimulates dilation of the cervix.

If, following several hours' straining, no feet are in evidence, then an examination is justified. This involves scrubbing the hands and arms thoroughly with soap and antiseptic. It is also necessary to clean the area of the cow's vulva so that dirt is not taken in with the arm. The hand should be pushed in steadily without bursting the water bag. If the head and feet are there, all should be well and more time needs to be allowed before giving further assistance.

Complications and Malpresentations

It is possible for cows (especially older ones) to suffer from milk fever during calving. One should therefore watch for an unsteady gait, padding of the hind feet and if in doubt, inject calcium boro-gluconate. Labour will normally then proceed within half an hour.

If assistance has to be given, e.g. the tongue is beginning to swell, it is best to attach calving ropes to the fetlocks of the calf and pull gently downwards towards the cow's hocks, but only when she is straining. If the head has to be roped, it should be placed behind the calf's ears. Sometimes the calf gets stuck at the hips; it is possible to help the situation by turning the cow on to the other side.

A common malpresentation is for the head to be tucked back. This problem can usually be corrected by gently pushing the brisket back and, by putting the hand under the jaw, carefully levering the head forward.

Posterior presentation is quite common, i.e. back legs first. This should be recognised by the feet of the calf being upside down, which then involves an examination to feel for the hocks and tail. In this position, once the hips are through, a firm steady pull is required as delay with this form of presentation will lead to a suffocated calf.

Head out and legs back is a complication which requires veterinary assistance, as does breech presentation.

Care of the Newborn Calf

Following birth, it is important to ensure that the calf starts to breathe, otherwise smart action should be taken. Several possibilities are available:

- Push a firm piece of straw into the nose
- Briskly rub the calf with a handful of straw
- Slap briskly on the side of the rib cage
- Throw a bucket of cold water over the calf's head
- Artificial respiration—laying the calf on its side and working the upper leg round in a circular motion;
- Kiss of life—blocking mouth and one nostril, blowing into the other whilst alternately pressing on the ribs to deflate the lungs.

To prevent infection through the navel it is necessary to dip this in iodine solution or to spray with an antiseptic aerosol.

The afterbirth is normally expelled some 12–24 hours after the calf, its retention leading to infection in the genital tract. If it is not

38. Spraying the navel of the new-born calf with antiseptic from an aerosol can. This is essential to prevent infection entering the body by this route.

expelled after three days veterinary attention should be given. On no account should herdsmen pull on the membranes as they may break and the portion remaining in the uterus will then cause inflammation and serious infection.

Early Lactation

It was stated above that the birth of the calf was a critical time for the cow. That is certainly true but in terms of the cow settling into lactation, achieving a high peak and total yield, the few days after calving are equally, if not more, critical. As calving is a natural process it should ideally involve minimal interference from the herdsman, but his role really does become important in the days that follow calving in terms of controlling feeding, milking and general management. Avoiding stress in general is a high priority which involves the provision of a clean dry bed, not allowing her to spend more than twenty-four hours with the calf, and moving her into a relatively small group of animals. (It may be worthwhile forming a special fresh-calved group.) One of the main aims at this time is to allow the cow to attain her potential appetite as quickly as possible, so that control of feeding and the avoidance of any digestive upsets are vital. She will require sufficient time to obtain her food and there should be no physical restriction to eating, such as inadequate pasture or lack of trough space. It is important to keep the rumen working well so that a supply of some 6 kg DM from good quality forage is required daily. Ideally this forage will be identical to that she was receiving pre-calving. Two kilogrammes per day of sugar beet pulp or brewers' grains, due to their high palatability and relatively slow rate of carbohydrate digestion, will also be a valuable part of the diet. The feeding of concentrates should be very carefully controlled, starting with 3 kg per day and building up over 10–14 days to a level of some 12 kg. Such detailed control should be possible in a cowshed or system where yokes or individually pro-grammed out of parlour feeds are available. It is much more difficult to control rations if the freshly calved cows have to join the main herd or a high-yield group where high levels of concentrates are being fed. Few problems should arise in complete diet fed situations as, due to the lower appetite of fresh-calvers, they will not be able to over-eat at this time. When concentrates are fed in the parlour, and especially at flat-rate levels, it is vital to identify these freshly calved animals at each milking and to reduce their feed of concentrates to the required level.

Once they are settled into lactation it is essential to meet their energy needs and to offset their continuing low appetite by offering a diet with a high proportion of concentrates. However, to avoid digestive upsets and the possibility of low butter fats, the concentrate:forage DM ratio should be in the region of 3:1 and should not be over 4:1.

Body condition score will drop by 1 to 1½ scores in the first 6–8 weeks of lactation associated with a loss of some 50 kg liveweight. Higher yielders have the ability to eat more food, but not sufficient to meet their increased needs, so they tend to lose more weight than lower yielders.

The important point for herdsmen (and owners) to note is that if this level of care and attention to detail is to be given to the freshly calved animals, and particularly to heifers, then adequate time has to be made available. Lower priority jobs will have to wait or ideally many of them will have been done before the calving season is in full swing. In herds which have a tight calving pattern the provision of extra help will be well justified at this critical time—as is often provided for a shepherd at lambing time.

The management of the cow during the breeding season will be dealt with in Chapter 11 covering such aspects as oestrus detection and optimum condition score. The role of the herdsman during this time is again a major one in ensuring that the diet continues to be adequate and that the cow suffers minimal stress.

Mid-Lactation

After conception is confirmed, few problems should arise in managing the cow through mid lactation to regain weight without falling in yield more than 2½ per cent per week. Appetite is now at its peak so that the full potential of background feeds should be exploited. The body condition of autumn calvers will need to be watched at pasture, especially during dry periods, and buffer feeds provided as required.

Chapter 6

MILKING

THIS IS another major part of the herdsman's job, one in which his skills and understanding of cow behaviour have a considerable influence on the overall success of a dairy enterprise. Cows which are handled quietly and are content at milking time let down their milk effectively, but if they are stressed their yields will be reduced and their full potential of breeding and feeding fails to be achieved.

Milking is undertaken by a wide range of people, from herd owners who rarely miss a milking and know every animal in great detail, to the occasional relief who is unfamiliar with the cows as well as the milking plant. In many situations one person undertakes the milking alone, whereas in other large installations two or even more milkers are involved, working the routine together. Many types of milking facility exist, ranging from cowsheds to various types of parlours including a few sophisticated rotary installations. The detailed routine varies somewhat from one type of parlour to another, but since the herringbone parlour is the most popular it is the routine for this type that will be described later in the chapter —but first, a word about the plant that is involved and some other aspects of stockmanship.

Milking Plant

A well-designed and efficiently installed plant is essential to good milking but regular and effective maintenance is equally important. Some of this maintenance, such as replacing worn rubbers, is part of the herdsman's job but people with specialised equipment need to be called in to undertake mechanical tests.

Some herdsmen work with fairly basic installations of considerable age but find that, if well maintained, they do the job satisfactorily. Some look with envy at the latest automated plants erected on show stands or on neighbouring farms, whereas others question

39. A modern Trigon milking parlour, which is a 3-sided parlour with herringbone stalls. This type of layout increases milking throughput by 10 per cent over a normal herringbone with the same number of units. The operator has a shorter length of time to wait for a slower milking cow due to the smaller batch size.

the potential problems of such sophisticated equipment. Herd owners contemplating the updating of a plant may consider utilising a temporary bail during the alterations, but will need to plan and control such an event in great detail. Cows and herdsmen are likely to suffer considerable stress in such circumstances, so that the changes should be undertaken as quickly as possible and ideally at a time when a high proportion of the herd is dry or in late lactation.

Herdsmen should discuss with herd owners from time to time (but not too frequently!) possible modifications to the plant which would improve efficiency. Minor expenditure may be all that is required to improve the working environment in the parlour, such as additional heat in winter or ventilation in summer. Remote control of gates in the dispersal race which divert animals for treatment is an example of a minimal investment which considerably reduces effort and stress on milkers. Where increased yields are being achieved or herd size expanded it may be feasible to incur larger capital sums, to increase the potential efficiency or throughput of the plant. In a two-stalls-per-unit parlour this can be undertaken by doubling up the milking units or by extending the length to incorporate additional units.

Good herdsmen take pride in their equipment as they do with their cows. The building is kept clean and tidy with effective internal and external cleaning of the equipment after each milking session. Such a situation should not be too difficult to achieve on farms where one person undertakes most of the milkings or on larger family farms where the 'boss' does some of the milking. It is perhaps not surprising that it tends to be more difficult to achieve in large herds where a number of milkers are involved, perhaps on a shift-system, with divided responsibility for cleanliness. Herd managers in these situations have a key task in setting appropriate standards and ensuring that every operator 'leaves the parlour as he would wish to find it'. Where a team of herdsmen is involved one member usually has some mechanical interest so that he can, with benefit, be given responsibility for routine maintenance tasks such as the weekly check of oil levels in the vacuum pumps, cleaning of the regulator or adjusting the belt drive on a refrigeration plant.

Milking Staff

It is a basic requirement of satisfactory milking performance that operators are able to concentrate on the job in hand, the milking and observation of the animals. They need, therefore, to have confidence in the equipment, be sure how it operates and know what to do if problems arise. Herd owners and established members of staff should have no difficulty in this area, but relief staff or new employees, especially in one-man operations, often feel vulnerable and do not settle or enjoy the task. Herd owners can help reduce such problems by issuing clear explanations and agreeing a procedure in case of problems. For this kind of reason alone, a telephone in the dairy is a most valuable resource not only to call for engineers but also for other perhaps more serious emergencies.

In parlours operated by more than one milker the opportunity to train young people is readily available. When two milkers regularly work together it is a considerable help if their personalities match and they enjoy one another's company. Such problems do not arise in a one-operator unit, as he has only himself to blame if he overlays by ten minutes one morning; and the dairy radio can always be tuned into the programme of his own choice.

Pre-Milking Tasks

There are several important preparatory tasks to be undertaken before the cows are brought into the collecting yard. This will, in most situations, apply more to the afternoon than to the morning in order that milking is completed before the transport tanker arrives. Some of these tasks can be undertaken the previous evening if an early start to morning milking is required.

Following milk collection the bulk tank should be cleaned and disinfected. Where automatic washers are fitted the tanker driver will have activated the washing cycle so that only the final rise may be required. Filters should be assembled and the milk supply pipe safely connected to the bulk tank, not forgetting the plug! At this time it is essential to check the temperature of the bulk tank, ensuring that the ice-bank has built up as it is a contractual requirement to cool supplies down to at least 4.5°C. Where in-line coolers are in use it is necessary to check their effectiveness at each milking by monitoring the temperature of the stored milk.

It is also advisable to prepare supplies of routine items used during milking, such as paper towels and teat-dip. Fly spray will also be needed on some farms although most now use long-acting compounds or fly-repellent ear-tags. It is also a considerable help to prepare a list of those animals which will need to be diverted after milking for AI or veterinary treatment.

Parlour and collecting yard floors and walls should be well wetted at this time, so that dung splashes are removed more easily.

The job of bringing the cows from pasture or from winter housing is a most important one in terms of stockmanship. The task should therefore not be rushed so that an adequate length of time needs to be allowed in order that it can be undertaken with care and, therefore, satisfaction.

Detecting cows in oestrus is one of these key tasks, as disturbance of the herd often encourages mounting behaviour. When large groups of cows are brought in from pasture, they usually string-out over some distance so that it is difficult to identify a bulling cow towards the front of the herd. Enlisting the help of a colleague for a few minutes to meet the herd as it approaches the buildings can be most beneficial. Signs of ill health such as lameness or bloat are often more easily spotted at this time. It is also a convenient time to check water supplies because if a number of cows are seen to be jostling around an empty trough an inadequate supply pipe or trough capacity will need to be corrected.

When bringing the herd from grazing it is necessary to check the

40. Checking the bulk tank before the commencement of milking.

41. An in-line milk cooler. This is a very efficient system of cooling large quantities of milk. After being cooled, the milk is stored in a stainless steel tank to await collection.

42. Removing a dung-pat from the end of a cubicle bed. This is a job which can often be undertaken while the cattle are moving out for milking.

amount of forage (grass or other forage crop) which has been eaten and evaluate the supply remaining, especially when a controlled system of grazing is in use such as paddocks or strip-grazing. If an electric fence has to be moved this may be the most convenient time to undertake the job, particularly if the cows can be herded into the track, so as not to interfere with the move. Cows should not be rushed along tracks but allowed to come at their own speed. Concrete surfaces should not be allowed to get too muddy especially on a farm with a stony soil, otherwise foot troubles will be more of a problem. The lame cow at the back of the herd which delays progress towards the collecting yard should be an encouragement towards more prompt or more frequent attention to feet!

When you bring animals from a cubicle or kennel house you will usually find that a number are lying down and it will take some time for these to get up, back out and move down the passageways. It may be possible to use this time to remove any dung pats from the ends of the cubicles with a fork or even with one's boot, but do not forget to keep an eye open for bulling cows. This is also the best time to check the consistency of the dung as an indicator of optimum diet formulation, and one should always be on the lookout for tell-tale deposits of mucus or 'whites' on the edge of the cubicles from animals which could have disorders of the breeding tract.

Milking Routine

Once the herd is in the collecting yard milking should proceed without undue delay so as to minimise the time that cows are away from food, water and rest. In some high-performance herds which are milked three times a day forage as well as a water trough is provided in the collecting yard to reduce this problem.

After switching on the vacuum pump it is necessary to check the gauge to ensure that the working vacuum level is reached quickly. If there is a delay, check for air leaks, possibly at an open drain tap or a misplaced jar lid. The sound of the pulsators should be regular and air should be entering the regulator and claw-piece airbleeds.

All should now be ready for letting in the first batch of cows which, in the herringbone, involves opening the backing gate. There is usually some pushing and jostling by those animals most keen to get into the parlour for their feed, when this is given, or to have their milk removed. Some cows are only content when allowed to enter one particular side of the parlour or even one specific place in the batch (usually first or last). Many herdsmen cope with such demands by individual animals whereas others suffer considerably from such hassle and herd owners would be wise to cull these troublesome 'beasts'.

Where feeding in the parlour is practised, the activation of the dispensing mechanism can usually be undertaken as the cows enter the stalls. In systems fitted with an electronic memory it is only necessary to press the appropriate cow number for each stall, whereas in other more basic types, after identifying the cow, the herdsman may have to refer to a chart before activating the feeders. The use of coloured tapes fitted to tails or numbered ankle straps can be a beneficial aid to operating a differential feeding system. In large herds, subdivided into groups by yield level or in herds operating a batch calving system, feeding is usually simplified as all cows in the batch receive the same level of concentrates.

Now, by working in a set pattern, that is, starting with either the back or front cow of the batch, the milking routine begins. It is usually worthwhile identifying a slow milker within the batch and dealing with her first, so as not to delay the dispersal of the rest of

43. Activating the dispensing mechanism for parlour feeding. The number of each cow in the batch has first been entered in stall order. Using the electronic memory which has earlier been programmed with appropriate feeding levels, each cow automatically receives her individual ration.

44. Teat washing from a warm water jet.

45. A herringbone parlour fitted with automatic cluster removers. Once the teat cups have been applied it should not be necessary for the operator to return to a cow. Note the in-line detectors for mastitis clots in the long milk tubes by the legs of the cows on the left of the picture.

the batch when they are completed milking. The first operation is to take foremilk from each teat ideally with a rubber-gloved hand either into a strip-cup or on to the 'clean' floor. Any animal showing signs of clinical mastitis with clots in the foremilk or one with any milk abnormalities should be marked on both rear flanks with aerosol spray and receive separate treatment, which will be outlined below.

Teats and any soiled parts of the udder are now washed, ideally with a warm water jet. This handling of the udder and teats stimulates milk let-down due to the release of a hormone into the bloodstream. After the teats have been dried with paper towels the teat-cups should be attached. Many animals have a short let-down period so that a delay of teat-cup application will reduce yield. Starting with the back teats, teat cups are attached by holding the cluster in one hand and using the other to lift each teat-cup in turn, guiding the teat into the cup with one's finger ends. Once the vacuum is holding the cups in position, milk begins to flow and one can then move to the next cow in line, repeating the same procedure. In two-stall-per-unit parlours the opposite batch is prepared while the first one is being milked, whereas in one-stall-per-unit parlours it is possible to let two batches in together and to work around the parlour as though with one big group. It is essential to keep an eye and ear open for a cup or even a cluster which may fall or be kicked off the udder.

In parlours fitted with semi-automatic cluster removers it will be necessary to return to each stall after the milk flow is in full stream to switch on the take-off mechanism. Where fully automatic cluster removers (ACR) are in use it should not be necessary to return to a cow once the teat-cups are applied. Equipment which automatically transfers milk from the holding jars has a short delay built into the sequence which allows the milker time to check milk quality and quantity visually, in case transfer needs to be postponed. It is necessary to observe that cows are milked out properly and if in doubt to feel the udder and draw the teats. In most circumstances the milker will know the expected level of yield and will note those animals which are 'down' at that milking.

In some parlours milk flows directly into the pipeline and no jars are fitted. With such installations milkers need to be most careful when checking foremilk and, unless milk meters are in regular use, no check on yield is possible.

Where ACRs are not fitted, clusters need to be removed as soon as the milk flow has finished. This involves detailed observation and frequent movement along the pit to deal with individual cows ready for cluster removal. The process involves closing off the vacuum, which prevents any dirt being sucked into the cups, then releasing the milk from the jars—if fitted. Teat cups should not be left on too long, as overmilking can damage udder tissue and it also encourages slow milking. If a cow is three-quartered (i.e. blind or dry in one quarter) it is necessary when attaching the cluster to double back the teat cup or fit a specially designed plug which cuts off the suction.

After teat-cup removal all cows should have their teats dipped or sprayed with disinfectant before being released. A number of highly automated dairies have a set of spray jets installed in the floor of the exit race which are activated by electronic eye as each cow passes through the gate. The milking routine is completed by opening the front gate, usually by remote control, which releases the batch. The opportunity should then be taken to wash away any dung from the standings, a bucket of water being more effective than a hose.

Quiet and confident handling of the animals should in most

situations prevent too much kicking, but problems often arise with newly calved heifers and occasionally with an animal with a sore teat. If help is available the problem can usually be overcome by holding up the tail into an almost vertical position, while at the same time scratching the rump. Tying a rope around the middle of the animal, just in front of the udder, is an alternative but many herdsmen now use a specially manufactured bar which fits with one spring-loaded end over the backbone and the other under the flank. Milkers should remember to remove the device before the cow leaves the parlour.

The potential throughput of a parlour varies according to the extent to which mechanical and automatic devices are used. In Table 6.1 a full manual routine for a herringbone of 1.2 mins/cow gives a potential throughput of 50 cow/manhour. In each subsequent column one such aid is added to show the effect on the work routine

Table 6.1   Effect of progressive mechanisation/automation on milking performance

Elements	Minutes per cow						
Let out cow	0.20	0.10	0.10	0.10	0.10	Auto	Auto
Let in and feed	0.25	0.15	0.05	0.05	0.05	0.05	Auto
Foremilk	0.10	0.10	0.10	0.10	0.10	0.10	0.10
Wash and dry udder	0.20	0.20	0.20	0.20	0.20	0.20	0.20
Attach cluster	0.20	0.20	0.20	0.20	0.20	0.20	0.20
Remove cluster	0.10	0.10	0.10	Auto	Auto	Auto	Auto
Disinfect teats	0.10	0.10	0.10	0.10	Auto	Auto	Auto
Miscellaneous	0.05	0.05	0.05	0.05	0.05	0.05	0.05
TOTAL	1.20	1.00	0.90	0.80	0.70	0.60	0.55
Cows per manhour	50	60	65	75	85	100	110

Mechanised gates/barriers

Semi-automatic feed dispensers

Automatic cluster removers

Automatic teat disinfection

Automatic cow exit

Automatic cow entry

Source: M.A.F.F.

and milking performance. Automatic cow exit can be obtained by linking the exit gate to the removal of the last cluster. As yet no automatic cow entry is in commercial production.

Post Milking Tasks

Where the arrangements permit, it is preferable to allow the herd to move directly back to pasture or to housing as they leave the parlour. If a public road has to be crossed or slurry scraped from buildings while the cows are out for milking it will be necessary to hold the animals in a dispersal area. The time spent here should be kept to a minimum, so that you may need to move cows before starting to wash down.

The equipment, stalls and building should then be washed. The first task is to brush the outsides of the clusters and jetters in a hot detergent–disinfectant solution. Worn or perished liners and rubber tubes should be looked for and replacement undertaken before the next milking. Where the acidified boiling water (ABW) system of equipment cleaning is in use ensure that at least 15 litres per unit of water at not less than 96°C is available. (The majority of water heaters are fitted with time-switches.) If circulation cleaning is used the water temperature needs to be 85°C, using the manufacturer's recommendations regarding the quantity of detergent–disinfectant. Stall work may need scrubbing to remove dung splashes. A high-pressure hose is well suited for walls, but for floors, a high-volume, low pressure supply is ideal. At least once each week air pipelines and the interceptor need to be washed and sanitised, and in areas of hard water the pipework must be descaled.

Mastitis

Returning to the cow with clinical infection of the udder: she should be milked out with care, ensuring that the milk is not transferred to the bulk tank. Some systems allow the jars to be emptied separately and in others mastitic cows are milked by diverting the supply from the jar into a milking bucket for use in calf feeding. The infected quarter should be stripped out by hand before antibiotic treatment is undertaken. Hands need to be disinfected before dealing with another cow.

It will often be practical to proceed with immediate treatment in a herd where consistent results can be expected from the 'proven' type of antibiotic in use. But where mastitis cases are rare or there are varying types of infection a 10 ml sample of the foremilk should be

46. Washing the collecting yard after milking.

47. Injecting a long-acting
antibiotic for the control of
mastitis during the dry period.

taken into a sterile bottle before treatment and sent without delay to a laboratory. The results will indicate what form of treatment is most likely to be effective, whether the infection is streptococci, staphylococci, coliform or mycoplasmas. A full course of treatment must be carried out and detailed records kept as an aid to culling decisions. A wall-mounted blackboard visible from the pit is a useful facility to record cows undergoing treatment. Data from this board is copied into the permanent records as and when convenient. Another such board in the parlour is of value to record cows in oestrus as well as those wanted for AI or other veterinary treatment.

The use of long-acting antibiotics during the dry period is another recommended procedure in mastitis control. At the last milking it is necessary to clean the teat end thoroughly with white spirit after milking, before infusing the antibiotic. The teats are then dipped in iodophor solution and the udder left to stock up.

Summer Mastitis

This is a serious type of infection first seen as a hardening of a quarter or stiffening of a teat and loss of appetite. It is thought to be mainly fly-borne and is usually found in dry cows or heifers in tree-lined, low-lying pastures. At least once a day careful inspection of dry cows is essential and if the infection is suspected the vet must be called for speedy treatment to avoid loss of the quarter or even death of the animal.

This attention to detail where disease or infection are concerned, coupled with patience and care for the animals in all stages of their daily round and a desire to see the appropriate plant operating efficiently at all times are the hallmarks of good stockmanship; and are nowhere more important than in the milking routine.

Chapter 7

HOUSING

A RANGE of well-designed buildings can be of considerable assistance to the herdsman in the efficient management of his cattle. When such a facility is provided fresh it normally represents a major capital investment, so to justify it enterprise output needs to be increased or costs reduced, or both achieved. Direct benefits can be attributed to housing, as for example the improved growth rate and feed efficiency of beef cattle which are yarded rather than outwintered. More often, the benefits are indirect, due to herdsmen being able to carry out their duties in a better environment. It would be possible, for example, to obtain a low Total Bacterial Count of milk from an outwintered dairy herd but this is much easier to achieve if the cattle are housed in well-managed facilities.

Cattle can in fact cope remarkably well outdoors with both cold and wet conditions, especially if they are well fed and can find some shelter from the wind. Food wastage, however, is common in out-wintered systems and there is also the problem of damage to the soil from poaching by cattle and the passage of feed vehicles.

Not all farm buildings provide an optimum environment for cattle, especially if overcrowding takes place. Bad ventilation contributes to lung disease, wet bedding to mastitis and slippery floors to split pelvic bones. The layout of the actual buildings within the farmyard can also have a marked effect on the efficiency of operation. The modern unit not only has regular movements of cattle around the buildings but there is also considerable vehicular traffic. A typical problem is the turning of tractors and forage boxes at the ends of buildings which have an inadequate width of concrete apron. In such situations damage is often caused to buildings and equipment as well as considerable frustration to drivers.

The majority of herdsmen have to work with an established set of buildings and little opportunity, at least in the short term, to influence major changes. They can, however, by using their skills of stockmanship make the best use of available resources to enhance animal comfort, health and performance. If and when building alterations or new facilities are being planned, the herdsman and herd owner should refer to specialised texts for detailed information on building design and layout. Although this chapter is written particularly to help the herdsman in his day-to-day work utilising existing buildings, herdsmen should nonetheless have much to contribute to the design team of new or updated facilities. They will be able to comment on the factors which will influence their performance as well as that of the animals. Typical requests will be for simple, reliable systems (e.g. slurry handling) and for attention to detail by builders in such areas as the finished surfaces of concrete floors.

Maintenance

On some large farms full-time maintenance staff are employed, so that herdsmen have a minor involvement in repairs and upkeep of buildings. On the majority of farms, however, as and when time permits, herdsmen enjoy involvement in this type of work. This is to be encouraged, so long as the animals are not neglected and the work is not of a highly specialised nature such as electrical wiring.

There is much relevance in the saying 'A stitch in time saves nine' because a broken cubicle rail can easily cause an injury requiring real veterinary stitches. As with machinery, building repairs can only be undertaken satisfactorily if an appropriate set of tools is available. Hammers, saws and some joinery tools will also be required for fencing work, so will usually be on hand. Other more specialist items for such tasks as plumbing or painting will need to be made available as and when required.

As with all a herdsman's jobs, it will help if he prepares a plan for the routine building maintenance. It will be necessary to avoid important times such as silage making, straw reception and of course calving time. Much of the maintenance work will be undertaken in the summer months when the buildings are not in use. The herdsmen on many farms find it helpful on occasions to walk around the farmyard with the owner or manager to discuss possible maintenance and improvements. A person who is not regularly involved in the routine work may notice more easily that facilities such as gate hinges and fasteners are deteriorating.

Before leaving the subject of maintenance and repairs, a few words of advice to employers of herdsmen—especially to those with one-man units. At all times, the number-one priority for herdsmen is the care of the cattle. There may be some times of the year when they have, or could make, time for maintenance jobs; this is good. There will, however, be other times when as well as the routine husbandry tasks other urgent problems arise with the herd. These may be accompanied by additional work on the buildings such as burst water pipes in winter blizzards. These are the occasions when management back-up in the form of providing additional help really does matter. Not only is the herdsman pleased to have the help but also to find to his satisfaction that the boss really does care about him and the cattle.

The remainder of this chapter deals with the herdsman's tasks which he undertakes in the various types of buildings.

Cowsheds

The cowshed, byre or shippon is still widely used on many smaller farms. Dairy cattle are milked and fed in the building as well as being tied in the stalls during the winter months. On most farms, cowsheds are old buildings of substantial construction which are difficult and expensive to modify or expand. Pipeline transfer of milk to the dairy has widely replaced the carrying of buckets but much hard work is still involved in feeding and manure handling. Some cowsheds which have a double row of stalls and central passage can be cleaned out with a tractor and scraper. Others have automatic manure scrapers fitted in the dung channel, moving material from the building and loading directly into a manure spreader. In each case, serious thought to the economic handling of materials can pay good dividends.

In order to produce clean milk the floor and stalls need to be frequently washed. The beds, however, need to be kept dry, which involves the herdsman in regular attention to the litter. Oestrus detection is more difficult than with loose housing so that it may be necessary to turn the animals out into an exercise yard to allow mounting to take place. The cowshed is a good environment in which to train young herdsmen because it allows easy identification, individual feeding and close contact with the animals.

Strawyards

This can be a satisfactory method of housing cattle especially in areas where straw is readily available and where farmyard manure is of value in an arable cropping system. It is an example of a facility which the herdsman has to use to the best of his ability. He may be

48. Cowshed housing.

able to overcome some design faults such as a large open side, without major expense, by fitting plastic netting. A large building can sometimes be subdivided to reduce competition between the animals, save straw and improve performance.

Stocking density in straw yards is a critical factor but one which should be under the control of the herdsman. Where 'surplus' animals cannot be found alternative accommodation some culling may be necessary or perhaps contract rearing arranged if the 'spares' are replacement heifers. A useful guide to stocking density for dairy cows is to allow six square metres per animal. The actual optimum will depend upon size of cow, quality of straw and particularly on the shape of the yard. Long narrow layouts are preferable to square yards, as there is less cow traffic between bed and feeding area. Ideally the feeding fence is separated from the bed by a concrete strip which can be scraped at intervals.

Straw requirements for bedding as well as for any feeding purposes need to be budgeted in good time. Early-combined straw is usually of higher quality than that harvested later, so the storage site should be prepared and staff be ready to receive and stack supplies. On dairy or beef farms which also produce cereals it is to the herdsman's benefit to co-operate as fully as possible with the arable staff, possibly helping with transport or bale loading.

Ideally, traditional-sized bales are stored alongside the yard where they are to be used, avoiding the need for a vehicle when bedding-up. Where it is necessary to build outside stacks it is worthwhile ensuring that effective covering takes place so as to minimise the number of wet and broken bales to be handled in the busy winter months.

When a tractor and trailer have to be used during the bedding-up operation, wherever possible this should be a two-man job, especially if it has to be undertaken with cattle in the yards. No herdsman should need to be reminded of the need to keep a sharp pen-knife to hand but perhaps more need encouraging to have a convenient receptacle available for used strings. Leaving strings on the ground is not only unsightly but it is also dangerous to man and beast and a considerable menace if they are allowed to become entangled in machinery. Where large bales are in use it is necessary for the handling vehicle to be available at times convenient to herdsmen and stock. This is another example of the need for planning and for good team spirit when a vehicle, widely used for general farm handling duties, is also essential to the herdsmen in the bedding-up of yards.

In most situations it will be necessary to clean out the yards at least once during the winter housing period. This is yet another task justifying careful planning and execution, so as to avoid damage to buildings and fittings or undue disturbance to stock. Water troughs at this time are liable to be damaged unless they are well located, and they will also need cleaning out after the operation. This is a job which is often neglected, but should not be: cattle do prefer to drink from clean troughs. The task is made easier if the ball-cock is tied up for a period before cleaning out so that the animals reduce the amount of water to be baled out.

Kennels and Cubicles

This is considered by the author to be the most satisfactory method of housing milking cows. It can also be well suited to heifer rearing—but it is not as suitable as straw yards for beef cattle or down-calvers. Although the capital cost of kennel housing is much less than general-purpose yards (with straw bedding or with cubicles) due to lighter construction, maintenance costs are much higher.

The ideal kennel or cubicle building should have no more than thirty cubicles on each side of the central dunging passage. Slurry removal from longer 'runs' is inefficient and unless several cross passages are available between feed and sleep, much bullying can take place. Cubicle length and width need to match cow size and the

49. A well-located water trough in a loose-housed building.

50. Cubicles with a good firm flat chalk base.

dunging passage should ideally be 2.40 m wide. For large Friesian/
Holstein animals the recommended size of cubicle is 2.20 m long and
1.20 m wide. There is again a limit as to what the herdsman can do
to overcome unsatisfactory design but considerable scope for efficient
management, especially of the beds. Well-cared-for beds improve
cow comfort, cleanliness of udders and minimise the evidence of
injuries. Where head rails are fitted, accurate adjustment reduces
the amount of dung falling on the beds, leads to less work as well as
to healthier udders. The base of cubicle beds needs to be firm, with
no holes allowed to form, thus reducing the risk of cows being
trapped under the bottom rail of the divisions. Rammed chalk, when
available, has proved to be a suitable base material but a thin
(75 mm) covering of concrete makes a suitable alternative.

New supplies of bedding need to be provided several times each
week. The type of bedding material used on a particular farm will
depend upon such factors as location (availability and price) and
method of handling slurry. Straw and sand are both good bedding
materials except where slurry is pumped. Machines are now available which finely chop straw from the bale and conveniently blow
the material onto the beds; slurry can then be pumped. Sawdust or
wood shavings are a good buy in some areas, although coliform
mastitis has been associated with the material from some types of
wood. Rubber and other specialist types of mats are used by a few
farmers as a method of reducing the need for bedding material. As
well as having a high capital cost, mats tend to provide a hard and
therefore relatively uncomfortable bed and if allowed to retain
moisture (under or above) create an ideal micro-climate for bacterial
growth.

Dunging passages need to be scraped at least once a day to
prevent animals having to spend too much time standing in slurry. In
the majority of situations a tractor and rear-mounted scraper is used
for this purpose—fitted with a replaceable rubber blade. Some
farmers, keen to replace labour with capital and at the same time
wanting to minimise disturbance to the cattle, utilise automatic
scrapers. Many such units are activated several times each day by a
time switch. A safety device is built into the system in case the
scraper blade meets an obstruction such as an incumbent cow. New-
born calves do not inactivate the mechanism, so that down-calvers
need to be moved out of such buildings in good time. The author is
very much against down-calvers being in any type of cubicle building, having been involved in retrieving newly born calves out of the
slurry!

Returning to the subject of tractor scraping, it is rather an

51. Bedding cubicles with straw.

52. Scraping slurry from the dunging passage of a cubicle house. Ideally this should take place twice a day to prevent animals having to spend too much time standing in slurry.

53. Pushing slurry into an above-ground store.

54. Training a cubicle-refuser. Such problem animals are rare if cubicle housing is used in the heifer-rearing stage.

unpleasant job on many farms especially where there are poor layouts, narrow doorways or uneven concrete surfaces. It is a job which has to be done day in and day out, so that care and attention need to be given to the tractor and scraper (see Chapter 13). Scraping should ideally take place when the animals are out of the building but they should not be moved solely for the scraping process. Considerable organisation is necessary, especially in larger herds, to ensure that the job is done at times convenient to cattle as well as to staff. On very large units it is almost a full-time job, at least for major parts of the day. A problem can arise if the task is regularly delegated to a junior herdsman or student, who will easily lose motivation unless he is given some additional responsibility. As the person is spending so much time near or with the cows, he needs to observe them continually, a process which is greatly affected by motivation.

When introduced to cubicles the majority of cattle soon take to the system—but a few do not and present a problem, or perhaps a challenge, to herdsmen. Frequently these cows lie in the dunging passage so that on arrival in the milking parlour they are a genuine problem. It is worthwhile encouraging them to use the cubicles by offering hay in the front of the beds. Some herdsmen claim success from tying offenders into the cubicles for several hours. Others pen the cubicle-refusers to one end of a passage and place concrete blocks in the slurry so that the animals cannot lie down. Persistent offenders are better sold to farms which use non-cubicle housing. It is a fact that heifers which spend some of the rearing period in cubicles are less prone to this problem.

Calving Areas

Cows calve in a wide range of situations from pasture to specialised maternity pens. If a cow is to calve in a cowshed it is wise to tie the animal's neck chain or rope to the stall with a thin piece of string in case she is down and slips back. The string will then break so that she is not choked. Traditional calving boxes have a considerable labour input so that on many farms calving takes place in a covered yard. The calving of healthy cows on a straw bed creates few problems and at least the cow is unlikely to slip and split her pelvic bones. Facilities to subdivide the area are essential for pedigree herds and worthwhile in other situations, so as to ensure accurate matching of offspring to dam.

An ideal layout is a narrow building, subdivided into pens, with gates which can be folded back for mechanical manure removal. Pen fronts are also formed by gates which carry a feed trough and water bowl. When opened both allow the cow to enter the pen and also block the remainder of the passage so that a cow can be moved single-handed. If the passageway in front of the pens is kept tidy, cows can be checked out of working hours in low shoes, perhaps on the way back from an evening out.

TV cameras are in use on some farms to save the herdsman time and effort in checking cows due to calve. Pictures from the pen are transmitted to a small screen in the lounge or bedroom.

Maintaining a clean, dry bed is essential in calving areas to minimise disease. At the first sign of infection it is essential to move the animals out and thoroughly clean and disinfect the area. During the non-calving months it is advisable to leave the facilities clean, disinfected and free of animals.

From all that has been written in this chapter it can be seen that farm buildings can be a very useful aid to herdsmen in carrying out their routine duties. Care and attention needs to be given not only to the upkeep of the facilities but also to ensuring an optimum environment for the stock.

Chapter 8

HERD HEALTH

HERD HEALTH is an area in which the stockman has a major influence upon the performance and productivity of the enterprise. Ideally all the animals in a herd would be free from disease but it is clearly accepted that such a situation is not likely. A recent survey carried out by the Ministry of Agriculture indicated that 43 per cent of the cattle in the national dairy herd of Great Britain are at any one time affected by disease. The stockman therefore has a vital role in observing the stock, identifying problems and, most of all, nursing the sick back to health by carefully implementing a control programme.

The diseases are seldom due to a single cause but rather to two or more factors, including housing conditions and the quality of nutrition, both of which are considerably influenced by stockmanship. Some health problems, together with their effects on performance, are obvious (for instance a lame cow). On the other hand many others such as sub-clinical mastitis can remain undetected for long periods of time.

The role of the veterinary profession in the detection and control of disease is well recognised and increasingly they are involving themselves with farmers and herdsmen in a team effort to promote herd health.

To achieve success with such programmes the team needs to meet at intervals to plan and assess performance.

They should arrange for:

- A set of management practices to be implemented which reduce the risk of disease, e.g. dry cow therapy or anthelmintic dosing programme.
- A recording system which pinpoints problem areas such as conception rates or youngstock weights.
- Regular veterinary visits especially at such critical times as the breeding season, e.g. pregnancy check and examination of cows not seen in oestrus forty days after calving.

100

55. Assisting the veterinary surgeon during his routine visit in the breeding season.

56. A well-stocked veterinary cupboard.

In this chapter we are concerned with the stockman's role in health promotion. It will be a considerable help if he understands the structure and function of the animal body and he is referred to appropriate textbooks for such details. His major role is to provide good husbandry so that the natural functioning of the animal is promoted by adequate feeding and water supply, healthy, comfortable surroundings and optimum group size. He needs to have adequate isolation facilities and comfortable areas to house sick cows when they are separated from the herd. He should be skilled in detecting the symptoms of ill health, this being achieved by training and experience. The role of older stockmen cannot be over-emphasised at this juncture in pointing out to newcomers the detailed characteristics, often rare, of specific health problems. Developments are taking place in training aids such as a dummy cow and calf which can provide invaluable experience in 'calving' which is such an important aspect of the herdsman's job.

He must develop a working routine so that the animals have a regular pattern to the day in milking and feeding, fitting in other tasks when possible. Although cows spend a considerable amount of time resting, especially at night, they actually sleep for only a few minutes, rarely stretching the head and neck out flat. It appears that cows need to remain vertical so as to allow cudding to function properly and to prevent bloat.

Signs of Health

At pasture an animal which wanders away from the herd may be in ill-health although when operating a set-stocked grazing system the herd does tend to be more scattered so that abnormal posture may be a better way of identifying a sick cow. She may be standing with a dejected appearance—ears and head held low and back arched, or, when lying down, the head either stretched out or turned towards the flank. In general, healthy animals are alert and if they are disturbed their response is reasonably quick. Cattle that are dull and listless should be watched for metabolic upsets, acute mastitis or in case they have a foreign body in the stomach wall.

The skin of a healthy animal should feel soft and pliable when handled over the last rib; a tight skin and lack of bloom are often associated with feverish conditions or digestive upsets. Healthy animals show lick marks on their skin and the tail and hindquarters are free of dung contamination. The dung of a healthy animal should

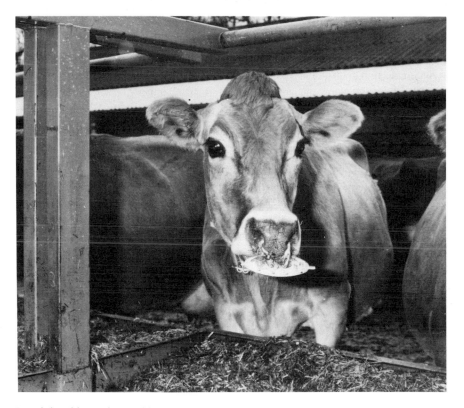

57. A healthy animal with moist nostrils, bright eyes and a shiny coat. Note the metal device in the nose to dissuade her from suckling other cows—a bad habit developed by a few animals.

be reasonably formed, depending upon diet and therefore on time of year. Very loose or very firm dung indicate a possible digestive upset. Strong-smelling or blood-stained dung or urine are abnormalities, as are coloured discharges from the genital organs.

The nostrils of a healthy animal are moist and free of mucus and the eyes should be bright. Sunken eyes and a staring look are often signs of the onset of disease.

Holding the ear of an animal is another good health check and it should feel 'warm'. Any fall in milk yield indicates some problem with the metabolic processes and further investigation has to take place. Any abnormality in the appearance or smell of the milk should be noted, as the smell of acetone (pear drops), for instance, indicates acetonaemia.

THREE HEALTH CHECKS

There are three particular checks on health which the stockman can use:

1. The Temperature

Amongst its many functions, the blood system of an animal regulates the heat of the body and in a healthy cow the temperature rarely varies more than one degree. The normal temperature of a cow is between 38 and 39°C and is taken by placing a clinical thermometer in the rectum.

The thermometer should first be well shaken to ensure that the mercury is down into the bulb, lubricated before insertion and then held slightly sideways in the rectum so that the bulb makes contact with the mucous membrane. Modern thermometers need only to be in position for some 30 seconds before they can be withdrawn and the reading taken.

Body temperature may be raised by bad housing, exertion of the animal—especially in hot weather—or by fear resulting from handling; but the main cause is fever.

Veterinary assistance should always be obtained if the temperature is over 39°C (103°F).

2. Pulse

This measures the rate at which the heart is beating and is taken by using the index finger of the right hand with gentle pressure on the artery underneath the tail. The normal rate is 45–50 per minute and a rate of over 80 is dangerous.

3. Respiration

This can easily be obtained by counting the number of chest movements occurring in one minute. The rates vary widely; animals breathe faster when young, following exercise, in hot weather and when frightened. The rate for a resting cow is 12–20 per minute.

Several major health problems are discussed elsewhere in this book, such as mastitis in Chapter 6 and infertility in Chapter 11, but some of the other more common ones will now be covered.

TACKLING COMMON TROUBLES AND DISORDERS ON THE FARM

Bleeding

To stop bleeding—make a pad of cotton wool, lint or even a handkerchief, moisten it in warm water and hold it firmly over the wound until a bandage can be applied to hold the pad.

A wound caused by a sharp instrument has relatively smooth edges and bleeds more profusely than one more jagged (as from barbed wire) which may take longer to heal although the bleeding is less profuse. If the blood tends to spurt out, then an artery has been damaged. Apply a tourniquet and call for immediate veterinary help, keeping the animal as quiet as possible.

Wounds

It is necessary to clip the hair from around the wound before cleaning up the area with a salt solution (two teaspoonsful of salt to one litre of boiled water). Damaged teats are a particular problem and may require stitching by the veterinary surgeon, especially if the teat-canal is penetrated. If wounds cannot conveniently be bandaged to keep them clean a dusting with sulphanilamide or antibiotic powder will minimise risk of infection.

Bruises

Bruises, including swollen joints, should be bathed with cold water followed a few hours later by warm water to stimulate blood circulation. If seen early by the veterinary surgeon, good control can usually be obtained. Good results are often obtained by using an Animalintex poultice.

Choking

If an animal has an obstruction in the gullet, caused for example by a potato, it is advisable to try and locate the obstruction by feeling along the length of the underside of the neck on the left side of the wind-pipe. Massage towards the throat will in most cases dislodge the obstruction but more difficult cases will require veterinary aid. *Never* use a broom handle or other instrument to push the obstruction down.

Eye Problems

Cattle sometimes get small pieces of chaff or other objects in their eyes causing inflammation and a discharge. It is usually a tricky job to remove the chaff by using the corner of a clean handerchief but slightly easier with a little eye ointment on the end of a small tube nozzle. If after several attempts the foreign body remains it is better to call the veterinary surgeon who will probably use local anaesthetic to aid removal.

An eye disease of increasing importance in cattle of all ages, but especially of yearlings, is New Forest Disease. Infected animals should be segregated, kept in the shade and protected from flies. Treatment with antibiotic ointment or drops may be necessary for several days although with mild cases immediate control often follows. It may be advisable to treat the whole group if several are showing signs of infection. Prevention may soon be possible following the development of effective vaccines.

Bloat

This is a condition caused by gases building up in the cow's rumen due to abnormal fermentation usually caused by lack of fibre in the diet. The left side of the animal becomes distended between the last rib and backbone or hip joint, breathing becomes laboured and prompt action is needed. It most commonly occurs at pasture so that with mild cases the animal can be walked quietly back to the buildings. A drench should then be carefully and slowly given of 28 ml of turpentine in 0.6 litres of boiled linseed oil (usually bought ready mixed) or a proprietary drench. Gentle massaging of the extended rumen may help in the expulsion of gas by belching. If, despite drenching, the bloat is getting worse it will be necessary to puncture the rumen wall. Traditionally a trocar and cannula has been used but a wide-bore syringe needle can be just as effective and leaves a smaller wound to heal. Care needs to be taken to locate the correct position for making the puncture. It is on the left side equidistant from the last rib, the point of the hip and the bottom of the spine (see Figure 10).

In emergencies any sharp knife can be used making sure that it penetrates at right angles to the skin. It is essential to be bold, and not dither, otherwise a dead animal will be the result.

This is a disorder of which prevention should be the key factor and one which can be markedly influenced by stockmanship. Avoid turning animals out on to young lush pastures unless they have received a good feed of fibrous material such as hay or straw. When bloat is experienced with strip-grazing it is a help to move the fence more than once a day so as to prevent the animals getting too hungry and then gorging themselves. The grazing of clover and lucerne is more likely to cause bloat than grasses so that it may be necessary to cut and wilt these crops before grazing in front of an electric fence.

Figure 10. The correct site for puncturing the rumen wall.

Displaced Abomasum

This is an increasingly recognised ailment which involves the fourth stomach or abomasum moving from its normal site on the lower right-hand side of the abdomen to the left side, under the rumen. The specific cause is not known but it is thought that the abomasum is displaced during the first six weeks of lactation. The cow is usually off her food so that the secondary symptoms of acetonaemia can confuse the diagnosis and veterinary advice is required. Surgical treatment has a high success rate and within twelve hours of the operation the cow is usually eating again, if carefully nursed.

Grass Staggers

This is a disorder caused by a deficiency of magnesium in the blood and is also known as Hypomagnesaemia. Magnesium helps to control muscle function and an animal with deficiency becomes unnaturally alert and its muscles twitch. It develops a staggering gait, may soon collapse and convulsions follow. Death can take place in a very short time unless treatment is given, i.e. a sub-cutaneous injection of magnesium solution. The disease is most likely to occur in spring and autumn when a sudden drop in temperature occurs and when herbage, especially from young leys, tends to be low in magnesium content. Preventative measures involve the supply of an adequate daily level of the mineral which is conveniently provided in 60 g of calcined magnesite mixed with molasses or rolled cereal or in purchased concentrates at specified inclusion rates.

Milk Fever

This is another metabolic disease caused by a low level of calcium and often magnesium in the blood. Known as Hypocalcaemia it usually occurs in the older cows (third calves and above) around the time of, or in the few days after, calving. The first symptoms are paddling with the feet followed by loss of control of the limbs. The cow is off her food, ears are cold to the touch but temperature is normal. An injection with calcium borogluconate under the skin of the neck should be given but it is always advisable to call for veterinary help as there may be additional deficiencies such as phosphorus or magnesium.

When a cow is down for several days she should be kept comfortable by liberal bedding, the brisket propped up and frequently turned to prevent pneumonia and pressure sores. As an aid to lifting a recumbent animal a large cylindrical bag is available on the market which can be placed under the cow then inflated with a 12 volt battery-powered air pump. Routine injection with calcium after calving may be a wise precautionary measure from the third calving onwards.

Acetonaemia

This is basically a digestive problem caused by the presence of excessive ketone bodies in the blood. These are the by-products resulting from mobilisation of body fat reserved which the cow has to undertake when the diet is low in readily available energy. The first sign of the disease is a drop in milk yield and a loss of appetite, usually of concentrates rather than roughages. The breath and milk of the cow have a characteristic sweet smell of acetone (pear drops). Temperature is normal but the cow is constipated and the dung is often slimy in appearance. Treatment involves the injection of simple sugars together with drenches of laxative or sodium propionate which helps the production of natural sugars in the rumen. It is helpful to feed carbohydrate-rich materials such as flaked maize.

This is another disease where prevention is particularly important and involves the feeding of a balanced and acceptable diet containing not only sufficient digestible energy and proteins but also adequate fibre. The skilled feeding of concentrates in the early weeks of lactation to avoid digestive upsets but still obtain peak yields should minimise the number of cases of acetonaemia.

Metabolic Profile Tests

Several of the nutritional disorders outlined above can be prevented or controlled more effectively by having information about the blood components of the cows in the herd.

Blood samples are normally taken from three groups within the herd: dry cattle, high- and low-yielders. Laboratory testing produces a comprehensive report on the levels of major blood components such as glucose, albumen (protein), calcium, phosphorus and magnesium. The interpretation involves comparison with standard figures and is best done in conjunction with the veterinary surgeon taking into account milk yield, body conditions and the composition of the diets being fed.

Lameness

This is a problem in most herds and a major one in many, so that all herdsmen need to be quick at recognising and treating lame animals. As well as serious loss in milk and meat production it can also lead to valuable breeding animals having to be culled prematurely. Sound feet and legs should be a factor in the selection of breeding stock although of more economic importance is the day-to-day management of the feet. Lameness can arise from uneven stony roads, muddy gateways or from allowing cows to spend too much time during the winter months in cubicle housing, standing with their feet covered in slurry. Regular paring and filing of feet will be less often needed if cattle walk on slightly roughened concrete to wear the sole of the foot naturally.

At the first sign of lameness the foot should be examined so that a crush or other suitable stalling needs to be available to enable the leg to be safely held to permit inspection. The foot should then be well washed and any foreign body, such as a nail or sharp stone, which could possibly have penetrated to the hard sole can then be removed with a hoof knife. The site should then be dressed with antiseptic and an injection of sulphanilamide given to prevent the formation of an abscess. If an abscess is suspected veterinary treatment may be necessary to cut away areas of the foot so as to release pus. The infected area must be bathed several days, and in between times dirt

58. Immobilising an animal for treatment of a rear foot.

59 A well-pared foot.

must be kept out and a bran poultice applied. The foot can be placed in a strong plastic bag and then be wrapped with a piece of sacking.

Foul in the foot is caused by a bacterial infection which can be picked up from dung or pasture but only via a wound or crack in the skin between the claws or around the bulbs of the heels. The 'foul' smell associated with this problem is due to the decomposition of tissue invaded by the germ. Control with sulpha drugs and antibiotics is recommended but prevention by a regular foot trimming and use of a foot bath is well worthwhile. A walk once a week through the bath containing 3 per cent solution of copper sulphate or 5 per cent formalin will destroy the germ and also harden and toughen the skin.

There is some considerable skill in making a good job of paring feet. The main message is to keep the toes as short and the heels as long as possible. A rasp is a very useful tool for this purpose but a sharp knife is also essential to ensure that a concave surface is left on the underside of the sole to minimise bruising of the softer tissue. Young herdsmen are encouraged to attend specialised training courses on treating lameness, as organised by the ATB.

Skin Problems

The skin of cattle occasionally becomes infected so that treatment by the stockman may be required. Lice cause unthriftiness, particularly in young cattle, but simple dusting along the neck and back line with louse powder is an effective control. Care needs to be taken when using some compounds, e.g. organo-phosphorus, in confined spaces, a face mask being essential to avoid inhaling the product.

Ringworm occurs most frequently in young calves and yearlings from fungal spores picked up from wooden fixtures such as penning or gateposts. Thorough annual cleaning and creosoting of wood is necessary as well as ensuring that the stock receive adequate nutrition. The availability of drugs which can be used as feed supplements or in aerosol form has considerably helped the treatment of this problem.

Allergies, photosensitisation and warts are other skin conditions which on occasions suddenly appear and often disappear for equally unexplained reasons.

Worms

The chapter on youngstock rearing will describe the problem of parasitic worms. Dosing of young cattle with anthelminitic is therefore recommended in early July especially if the animals have been grazing old pasture or a sward grazed by cattle in the previous year. Dosing of adult cows for worms may be justified in some situations, and is generally done immediately after calving.

AILMENTS CONTROLLED BY LEGISLATION

To conclude this chapter, there are a number of health problems which are in some measure controlled by legislation. They are:

Warble Fly

Widespread control treatment by farmers over recent years has drastically reduced this problem so that 'Gadding' cattle due to warble fly are fortunately very rare.

Legislation requires that any cattle found to be infested are reported to the Divisional Veterinary Officer. Movement is then restricted until all cattle on the farm over twelve weeks of age have been treated using a liquid systemic dressing.

Sudden Death

Anthrax, which is a notifiable disease, should always be considered following sudden death. The carcase of such an animal is highly dangerous so that it should not be interfered with or disposed of until veterinary advice is taken. Sudden death may be caused by poisoning or by various bacteria but in all cases veterinary advice should be sought at once. Poisoning can be caused by plants such as yew or hemlock or by chemicals such as sheep dip and lead paint.

'Blackleg' can cause sudden death particularly in young cattle at pasture so that vulnerable animals should be vaccinated by the veterinary surgeon before turn-out in the spring.

Foot-and-Mouth Disease

If this disease is suspected, it is preferable to notify the police or Area Veterinary Officer rather than to call one's own veterinary surgeon who in a positive case would then be prevented from working for a considerable period of time. The symptoms are lameness, a sudden drop in milk, high temperature and loss of appetite. Later there is dribbling from the mouth and blisters form on the tongue, gums, muzzle, teats and feet.

Tuberculosis

Another notifiable disease but now very rare in the United Kingdom.

Enzootic Bovine Leukosis

An infectious virus disease present in a small number of imported cattle. It can be identified by laboratory examination of a blood sample and the Divisional Veterinary Officer should be notified.

It is evident from reading this chapter that there is a very wide range of potential dangers to herd health. Nevertheless, the alert, confident and capable stockman with the help of his veterinary surgeon minimises these dangers so that they are the exception rather than the rule.

Chapter 9

CALF REARING

AFTER BIRTH it is preferable to leave the calf with its dam for 24–36 hours to enable it to suck the first milk, called colostrum. On some farms, for convenience, the calf is left with the cow for up to four days but this practice can create considerable stress to both at separation. Calves which are to be reared on the bucket-feeding system learn to drink much more easily when removed from the cow at the earlier time.

The need for satisfactory suckling in the early hours of life cannot be over-emphasised. If the calf fails to suckle unaided it should be guided to the teats and encouraged to suck. It may be necessary to insert a teat into the mouth of the calf and draw milk. If, following such assistance, the calf still does not suckle it should be given hand-milked colostrum from a bottle.

Colostrum

Colostrum or 'beestings' as it is often called, has a high nutritive value. It is particularly rich in protein and is a laxative which helps to remove the dung which has accumulated in the gut before birth. The protective substances contained in colostrum are most important as they give the calf immunity from the bacteria which are inevitably present on the farm. These are known as immunoglobulins or antibodies and are carried in the fat globules of the colostrum. Research work has shown that the rate of absorption of these immunoglobulins from the digestive tract into the bloodstream peaks some three to six hours after birth. This is the reason why early satisfactory suckling is so advantageous.

Following separation from the cow, whole milk should be fed to the calf twice a day for the first week of life. Surplus colostrum from freshly calved cows can be stored by deep freezing or by allowing it to go sour before subsequent feeding to bucket-reared calves. Some

114

herdsmen consider the handling of sour colostrum to be rather unpleasant but calves reared on it appear to thrive satisfactorily. A simple laboratory test has been developed to indicate the level of antibodies in the blood, using a sample from the calf. It is known as the zinc sulphate turbidity test and is a most useful aid to the rearer of bought-in calves.

Housing

Following removal from the dam the calf should ideally be placed in a clean, dry, well-bedded individual pen. This should be located in a well-ventilated but draught-free building maintained at a temperature of 8–10°C. The relative humidity of the air should be kept to low levels by avoiding the washing of buckets or using excess water within the actual calf house. In a large calf-rearing enterprise, operating for an extended period throughout the year, it is preferable to have several smaller calf houses rather than one large building. This enables an all-in all-out system to operate with cleaning and disinfection between use, which helps to control disease.

The use of slatted boarding under the bedding of the pens can be a help in reducing straw usage by improving the drainage, but it does increase the time required to wash and disinfect the facilities after use.

Natural ventilation is adequate for the majority of calf buildings, providing that condensation does not take place in winter conditions. Power-operated systems using thermostatically controlled fans are installed in some units in order to achieve a satisfactory environment. Air movements need to be maintained even in cold weather. The ideal system is one which automatically heats the incoming air when the outside temperature drops to a set level.

METHODS OF CALF REARING

1. Single Suckling

This is the most simple system of rearing in which the calf suckles its own mother from birth to weaning at nine to ten months of age. It is an expensive system, especially when using high-grade land, and is only justified for the production of first-class animals for beef or for valuable pedigree animals. It is, however, well suited to low-output upland holdings or to arable farms where the cows can utilise crop by-products. The cow is usually of a pure beef breed or is beef × dairy

60. Having suckled colostrum from its dam, this newborn calf has been placed in a clean, dry individual pen for bucket rearing.

61. Friesian calves with concentrates and water to encourage early rumen development.

with sufficient milk production for the one calf. When crossed with a beef bull such cows produce high-quality animals. Natural service is the usual system of rebreeding suckler cows, although prostaglandins are being increasingly used to synchronise oestrus so that AI can be more conveniently used.

Magnesium deficiency (grass tetany) is often a problem with the cows and vitamin E deficiency (muscular dystrophy) is common in suckled calves. Occasionally a problem can occur when the calf is unable to consume all the available milk, leading to mastitis.

Careful observation is required by the herdsman and this is much easier in a small herd compared with one where large numbers of cows and calves are running together.

2. Multiple Suckling

This system involves suckling several batches of calves on a cow during the lactation. The number reared will depend upon the milk yield of the cow and the number of weeks each batch is allowed to suckle. Using a cow of a dairy breed and allowing eight to ten weeks suckling of each batch, then fifteen to twenty can be reared in one lactation. Although used on a few farms to rear dairy heifer calves, it is most commonly used for the rearing of purchased calves for a beef enterprise. The ideal cow is a high-yielding dairy type of reduced value due to some conformation defect which causes milking problems in a modern parlour system. Wide teat placement or loss of one quarter are common reasons for culling from the dairy herd so that if the animal has a quiet temperament she would be ideal for rearing calves. It is often difficult to identify oestrus so that the cow fails to re-breed and is culled.

Considerable patience is required by the herdsman when persuading the cow to take to her calves, especially to a new batch following weaning of the previous group. Some hand milking may be necessary for several days at this time until the new calves are able to consume all the available milk. With some cows this can be a rather challenging task and a good test of stockmanship! It is a help if the cow can be tied when mothering-on new calves and it may be necessary to tie a rope around the abdomen in front of the udder, to stop kicking. It is advisable not to include the cow's own calf in the first batch, moving it to another cow once it has obtained colostrum.

It is a system well suited to a small-scale rearing enterprise producing quality weaned calves at periodic times throughout the year. Health problems tend to be similar to those of single suckling, although the incidence of nutritional scouring is often higher.

3. Bucket Rearing

This is the most widely used method of rearing calves, both for beef production and for dairy heifer replacements. It is a suitable system for rearing home-produced and purchased calves but in the latter case much more care is required to minimise disease problems. The system is based upon the use of purchased powdered milk replacers which are mixed with the appropriate amount of water and fed from seven to ten days, until weaning usually at five weeks of age. A wide range of replacers are on the market from dried milk products with added fat, sugars and minerals to lower-cost vegetable-based powders. All tend to be rather expensive so that the aim is to limit the period of liquid feeding, commensurate with a good growth rate.

Individual penning is ideal so that the intake of each animal can be monitored and controlled. Contaminated buckets are an obvious. source of infection, so that thorough cleaning and disinfection are essential at least once a day.

It is often necessary to spend time teaching calves to drink from a bucket, especially those which have been suckling their dam for several days. It may be necessary to get into the pen, stand alongside the calf and back it into a corner. Hold the bucket in one hand, put the other over the calf's head, and encourage it to suckle the fingers. Once suckling, the head can be lowered into the bucket, gradually removing the fingers as the calf takes to the milk. Holding the bucket at knee height allows the calf to drink more readily and hopefully the bucket can soon be placed in the holder on the front of the pen.

Traditionally, milk replacer is fed by bucket twice a day at or around blood temperature. There are, however, some advantages in raising the temperature of the water used for mixing to 40–42°C. The calf then drinks more readily as the flow of saliva is stimulated and the rennet, which is present in the stomach for clotting the milk, is most active at this higher temperature. Consumption should be allowed to build up over the first week until full intake which for a Friesian calf is 4 litres per day in two equal feeds using 125 g/litre reconstitution. Actual mixing instructions will be printed on the bag or on an attached leaflet and these should be accurately followed.

Once-a-day milk feeding is an alternative and well-proven system. It involves feeding the same daily level of powder but in a more concentrated form—using 150 g/litre of water. The system is well suited to farms where limited labour is available. Milk feeding on these farms can conveniently be undertaken between milkings, usually in the late morning when other chores have been completed.

62. A batch of calves on an all-in/all-out early weaning system. The building will be cleared, disinfected and rested between batches.

63. Ad-lib feeding of acidified cold milk. Some time is usually involved teaching young calves to suck the artificial teat.

Another alternative involves mixing the milk powder with cold water so that the liquid is fed at tap temperature. Calves appear to thrive quite well although they sometimes shiver after a feed; fuel costs are reduced and it is a system well suited to locations where electricity for water heating is unavailable.

4. Ad-Lib Systems

Numerous ad-lib systems of feeding milk replacer have been developed in recent years, particularly to save labour. They involve either various types of dispensing machines or simple storage vessels connected to a series of rubber teats. Rapid growth rates can be achieved from these systems due to the high levels of milk intake coupled with the little-and-often suckling. Animals can be group-housed thus saving on the cost of individual penning. Some calves need to be trained to use the teat and odd ones that do not take to the system may need to be taken out and bucket fed. Nutritional scours may be seen in the first week; most will clear without treatment but any that have blood-stained faeces should be carefully watched and veterinary advice obtained. Stockmanship is therefore of major importance in ad-lib systems, particularly avoiding the temptation to reduce the frequency of inspection.

Machines can be programmed to provide continuous supplies of milk in the early weeks of rearing, followed by intermittent supplies as the time of weaning approaches. Thorough daily cleaning is essential, especially of the parts which are in contact with the liquid milk.

Specially formulated acidified replacers have been developed which stay fresh for up to three days. They are ideal for use with bulk storage vessels (150–200 litres) as the powder mixes easily with cold water. Some tanks have the teats directly connected so they need to be located in the pen and are therefore difficult to service. Others, more convenient to service, are located in the passageway and supply milk via plastic tubes to a bank of teats attached to the pen front.

With all milk feeding systems clean water should be provided from the first week of life. Early-weaning concentrates should also be offered from the second week, these being commonly purchased as small pellets or 'pencils' or a home-mixed ration. The main requirements of this feed are a high nutritional value coupled with palatability. This can be achieved by including such constituents as dried skim milk, molasses or sugar beet pulp. A home mix can be made up as follows:

	%
Flaked maize	50
Crushed oats	20
Fish meal	10
Linseed cake	10
Dried skim milk	10
Plus minerals and vitamins A and D	

A little soft hay is considered by many rearers to aid rumen development but others obtain satisfactory performance by allowing calves to take fibre by nibbling straw following daily bedding.

Veal Production

Demand for veal in the United Kingdom is small (0.2 kg per person per year), and the majority is utilised by the catering trade. The aim is to produce 'white veal' by feeding only high-fat milk replacer diets from birth to slaughter at 14–16 weeks of age. It has until recently been common practice to rear veal calves in individual slatted-floored crates without bedding. The crates were located in well-ventilated buildings, and to keep the animals as quiet as possible and reduce fly problems semi-darkness was maintained. As the animals approached slaughter weight (225 kg) they were too large to turn around in the crates so could only move up and down. These restrictions, together with the complete lack of fibre in the diet, has caused considerable public concern so that alternative rearing methods are being developed.

One feed company keeps small groups of calves on a straw-bedding system and with ad-lib milk feeding produces veal of most acceptable quality.

Husbandry Tasks

One of the earlier husbandry tasks to be performed is the *removal of supernumerary teats* from heifer calves being reared for replacements. This should be undertaken in the first month of life (by law before three months), cutting off any surplus teats by using a pair of sterilised sharp curved surgical scissors. Usually there is little bleeding and it is sufficient to hold a small pad of cotton wool over the site for a minute or so.

Castration of young bull calves destined for beef is necessary unless an intensive feeding system is in operation which can utilise entire males. A common and reliable method of bloodless castration involves the use of rubber rings which by law are restricted in application to the first week of life. The ring is first fitted on to the hooks of the special pliers, then by squeezing the handles together the ring is stretched enough to fit over the scrotum. Both testicles need to pass through the ring and on gently releasing the pliers care has to be taken to avoid nipping the small teats.

An alternative bloodless method is to crush the cord of each testis using a Burdizzo Castrator. Each cord in turn has to be pulled to the side of the scrotum, placed in the jaws of the tool and crushed by closing the handles. To ensure success, each cord should be crushed in two different places.

Castration with a knife is very effective but is really a job for the veterinary surgeon. He has by law to undertake the task if the animal is over two months of age and an anaesthetic has to be used.

Disbudding is another herdsman's task on farms where calves of horned breeds are reared. The law in this case requires anaesthetic to be used unless the task is done by chemical cauterisation in the first week (not a very reliable method). The most satisfactory method is to cauterise the bud with a special hot iron heated by electricity or by the flame from a bottled gas supply. This task should ideally be undertaken before three weeks of age but there does need to be sufficient growth of horn to hold the cauteriser in position. Local anaesthetic is first given to freeze the nerves running to the horn buds. The site of the injection is in the soft tissue just below the ridge of bone half-way between the eye and the horn bud. The manufacturer's instructions should be followed regarding quantity to use and the injection should be made fairly deep at right angles to the ridge. Hair round the horns should be clipped back to expose the buds during the five to ten minutes required for the anaesthetic to take effect. The effect lasts for up to one hour so that all the group to be disbudded can be injected together. With an assistant firmly holding the head of the calf, the cauteriser is then held on the bud for up to ten seconds, and with a twisting action the bud is pushed out by pressing firmly inwards and downwards with the edge of the cauteriser.

Care should always be taken to avoid fire risks when using gas appliances or electrical extension leads. Keep them dry and out of the reach of calves.

64. Injecting anaesthetic prior to disbudding. The site of the injection is in the soft tissue below the ridge of bone halfway between the eye and the horn bud.

65. Cauterising the horn bud with a gas-heated iron. It is held on the bud for up to ten seconds and with a twisting action the bud is then pushed out.

66. The horn bud cleanly removed.

SOME CALF REARING PROBLEMS

Gastro-enteritis

This is a common disease of young calves. Scouring can be caused by over-feeding but is often associated with infectious bacteria, viruses or mycoplasma organisms. Particularly virulent bacteria are E. coli and strains of Salmonella which can cause serious losses.

Respiratory Infections

Calves housed in badly ventilated buildings are liable to suffer from respiratory disease, often called calf pneumonia. The cause is complex, it does involve the environment but also involves micro-organisms and stress. Improved ventilation is an obvious control method but vaccines are also being developed against this disease.

Abnormal Calves

Occasionally calves are born with hereditary defects such as dropsy or bulldog calves. One such abnormality is a freemartin heifer which, as twin to a bull calf, suffers an arrest of sexual development before birth. The AI organisations appreciate notification of abnormalities so that if there is sufficient occurrence future users of the semen can have appropriate warning.

Chapter 10

HERD REPLACEMENTS

THE AVERAGE productive life of a United Kingdom dairy cow is rather less than five lactations so that for a 100-cow herd some twenty heifer replacements are required each year. In suckler beef herds the stress of production is much less than in dairying so that a typical culling rate is 10–15 per cent. In dairy herds a regular annual replacement usually takes place whereas in a beef herd a group of heifers are more commonly reared or purchased intermittently.

The actual replacement rate in a particular dairy herd is affected by many factors. Market prices of cull cows and replacement heifers often influence replacement decisions especially in a business where the cash situation is a critical factor. The retention of a number of 'borderline' culls for an additional lactation and either selling the proposed replacement heifers or avoiding purchase, has a marked positive effect on cash flow. It must be stressed, however, that such a policy cannot be maintained.

In many herds high levels of enforced culling takes place, as disease problems override any decisions by management. Chronic lameness and mastitis are typical problems but the most common one is failure to conceive. This is another major area in which the herdsman is able to make his mark on the profitability of the herd by maintaining a high herd health status and so reducing the level of enforced culling. Unless exceptionally high numbers of replacements are introduced into herds with health management problems, few cows can be culled for low yield, milk quality, conformation faults or behavioural problems. In some circumstances, low-yielders can be justifiably retained if they breed regularly and have no other health problems. Although in some herds three-quartered cows are often retained, any animals which cause persistent problems to herdsmen, such as a continual kicker, should go smartly 'down the road'.

Costings from farm surveys clearly demonstrate that the returns to

125

be obtained from using resources to rear replacements are lower than for milk production, so that high replacement rates create an additional cost burden. The rearing of an above-average requirement of replacements is often claimed to be justifiable when off-land is used which it would not be convenient to use for other purposes. Farmers who do this need to be aware that other resources such as labour, cash and their own management are involved in the enterprise, as well as the land. Now that the risk of diseases such as brucellosis has been eliminated, such inconvenient fields might be more profitably let to other producers.

In a herd with a normal culling rate, and allowing for the birth of bull calves, half the herd will be required for breeding heifer calves for replacements. It is an interesting point for discussion as to which of the cows should be used for this purpose. In a pedigree herd, particularly in one selling surplus animals as breeding stock, the highest yielders and cows with superior conformation will normally be used regardless of date of calving. In strictly commercial milk-producing herds, especially in those with a tight calving pattern, it will be the first half of the herd to calve which should breed the replacements. National Milk Records now rank animals in the herd in terms of their relative breeding merit but, again, this does not take calving date into account.

Rearing Policies

There are three main alternative methods of obtaining replacement stock, namely to purchase, to home rear, and to employ a contract rearer.

The purchase of down-calving heifers is the usual method for a 'flying' herd but is also used periodically by a wide range of dairy and beef farmers. Animals can be obtained of a required size, age, calving date and particularly of required conformation. Although considerable lump sums of cash are required (with some risk as to the actual amount), money is not tied up in a rearing enterprise. Neither land nor management time is involved and in the present situation disease risk is minimal, especially if none of the heifers has calved before arrival. Buying 'on the farm' rather than in markets reduces the disease risk, although travelling to dispersal sales can be expensive in time and money, and the cost of transporting the animals home is also a factor to be considered. Some farmers and their herdsmen get considerable satisfaction from studying performance data of sires and dams as well as from attending the sales to select their replacements.

Many more prefer to rear their own heifers, to avoid disease risk and to have the pleasure of seeing the result of planned matings develop into full members of the herd. As mentioned above, considerable resources are involved and few producers are aware of the actual full cost of producing a home-bred downcalver.

Contract rearing, although not as yet widely practised in the industry, does offer considerable advantages to the dairy farmer who is short of land and perhaps labour. He is able to have the calves reared from his own cows at a known cost as well as with minimal involvement of his time. Disease risks depend upon the number of other producers the rearer is dealing with and quality of downcalver, to a large extent reflecting the skill of the rearer. The system offers the rearer a regular income with reduced capital outlay. It is well suited to a grassland farm which has perhaps discontinued in milk production or to an arable farm as a means of utilising rotational grassland without capital injection. A carefully worded contract needs to be drawn up, to include such detail as which party covers the cost of veterinary bills and anthelmintics.

In beef herds the source of replacements is somewhat different as many suckler cows are commonly cross-bred animals, so are rarely born in the home herd. Beef × dairy heifers are very popular especially the Hereford × Friesian which can be readily purchased at almost any stage from the young calf to the downcalver. Other more specialised crosses such as the Blue–Grey can be purchased or produced at home by the herd having a few pure Galloway cows which are served or inseminated with a White Shorthorn bull.

Age at Calving

The next factor to be considered by farmers rearing heifers or having them contract reared is the optimum age of calving. It is a most important factor because not only does it determine the resources involved in rearing, but it also affects the calving pattern and long-term production of the herd. There is considerable evidence to indicate that so long as satisfactory growth takes place, heifers calving at two years of age rather than 2½ or three years have a higher lifetime production. Although the yield is rather less in the first lactation, younger-calved heifers are, on average, retained in the herd for a longer period of time, so producing more milk. Many farmers and their stockmen have reservations about early calving but tend to confuse the interaction between age and weight, knowing that small heifers usually have unsatisfactory performance.

Research findings show that it is unsatisfactory to grow heifers on a very high plane of nutrition to calve at less than 22 months; dystokia being a major problem. With good feeding and management it is a realistic target to produce a heifer weighing 500 kg just before calving at two years of age.

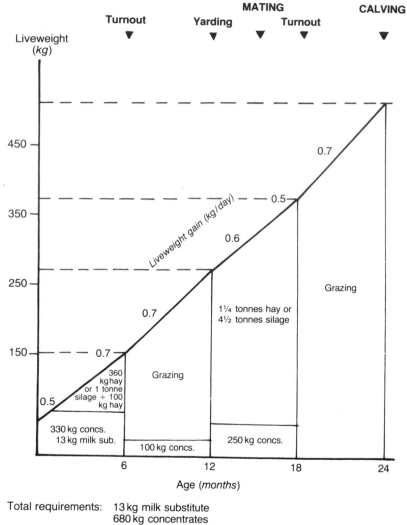

Figure 11. Targets for autumn-born calves to calve at two years.

An associated factor to be considered along with the age at calving is the calving pattern of the herd that the heifers are to join. If there is a tight seasonal pattern (most animals calving within two months) as in a few dairy and beef herds, then to maintain the pattern, heifers need to calve either at two or three years old. In herds with a spread of calving dates there is a wider range of options for age at calving, especially if no attempt is being made to tighten the pattern. The calf- and heifer-rearing programmes are complicated on farms which retain heifer calves over an extended period. Unless numerous groups can be formed competition is a problem in a group with widely varying sizes of heifers.

Owners of some high-yielding herds prefer to calve their heifers as a group, several weeks before the main herd, so as to allow staff more time to settle the heifers into lactation satisfactorily and so lift peak yield as well as total lactation production. Some delay is then allowed in first service which, together with a rather longer dry period, allows the heifers to continue growing prior to their second calvings.

Whichever of the rearing systems is adopted, it is advisable to have a definite plan for heifers to calve in a given month. Once this date has been decided, growth rates to achieve that target can be assessed. The major task is then to ensure that nutrition, health and environment are maintained at a satisfactory level for adequate growth. Target growth rates need to be achieved at every stage of rearing, as the need for accelerated growth rates in subsequent periods is not only expensive (if only because of the need for concentrates) but it may also lead to over-fat heifers and to more dystokia.

Yet again there is the need for first-class stockmanship to observe and monitor performance so that satisfactory growth is obtained throughout rearing. The role of management is also important in providing the necessary resources, particularly quality forages, adequate housing, and handling facilities. Figure 11 indicates the target liveweight gains for the heifer calving in the autumn at two years old. This is the most common type of heifer being reared and so will be used as an example in describing the detail of a rearing system in the remainder of this chapter.

Calf Rearing Stage

In the case of the autumn-born calf, targets are more easily achieved with the earlier-born arrival, that is if they are born in September rather than December. Older calves at pasture during their first grazing season have a higher appetite and so can make better use of grass.

During the calf stage, an early-weaning system is ideal, using bucket rearing or an ad-lib system of milk feeding as described in the last chapter.

First Grazing Season

Few difficulties should arise until turnout which is ideally at six months of age and 150 kg weight, but then the major challenge to herdsmen begins. Where husk is a problem vaccination will need to take place some six and two weeks before turnout. A good leafy sward will maintain growth rate but over-lush or stemmy grass is less satisfactory. Concentrate supplementation in the early weeks at grass (up to 1.5 kg per day) will be beneficial and may need to be reintroduced in very dry periods but particularly at the end of the season as forage quality declines. Newly sown leys provide an ideal feed, being free from parasitic worms but older leys or permanent grass have often to be used. Grazing management is therefore most important in such situations allowing the young heifers to satisfy their appetites with nutritious grass. A move to 'clean' aftermath in late July is of benefit, possibly after second-cut silage, or even to a

67. Young calves at grass.

pasture grazed earlier by another class of animal, such as ewes and lambs. Dosing with anthelmintics in July has been to date the recommendation for worm control but following the introduction of the long-acting sustained-release bolus, these are better 'put' in the rumen before turn-out.

Set stocking and paddock grazing are widely used systems for calf grazing but good results have also been obtained with the leader–follower system. This involves the young calves grazing one paddock ahead of a group of older heifers providing the calves with selective grazing. Stocking rates are dependent upon the level of nitrogen fertiliser applied so that if a high level of 300 kg per hectare is used, two livestock units (two young heifers plus two in-calf animals) can be carried per hectare.

Regular weighing of the animals when at pasture is recommended so long as the handling facilities are close by, so minimising the disturbance which is created. A move to better grazing or even concentrate feeding may be required if target growth is not being achieved.

Yearlings—Housing and Service

The changeover period to housing at the end of the first grazing season is another critical one for avoiding any check to growth. Introducing concentrate as well as a little hay or silage at pasture will help the rumen flora to adjust to the change of diet.

Target weight at service for a Friesian heifer is 330 kg and for early autumn calving the service has to take place in December which does not allow much time for recovery from any loss of weight on housing. High-quality conserved forages are essential at this time together with concentrates at least until service.

The choice of bull to use on heifers is influenced by the need to avoid difficult calvings but at the same time to produce a valuable beef cross calf or even a replacement heifer—if the herd is being expanded. Large beef breeds should be avoided. The Aberdeen Angus will produce small calves with the minimal dystokia but lower growth potential than the Hereford which is widely used for this purpose. Natural service prevents the need for oestrus detection which can be a difficult task in the winter months, especially in isolated locations without electricity supply.

The use of prostaglandins has proved to be a convenient method of bringing a group of heifers into oestrus together so that AI can be used without oestrus detection. Satisfactory conception rates appear to be achieved when the animals are in a rising plane of nutrition at

68. Yearling heifers at the drinking trough. Note the glossy coats indicating good health and growth.

69. Yarded heifers.

service time. The usual system is to arrange for the vet to inject prostaglandin intramuscularly on two occasions with an eleven-day interval. Insemination is then carried out at both seventy-two and ninety hours (days three and four) after the second injection. Special arrangements need to be made with the AI centre to book the necessary semen but more especially the inseminator time. Some saving in cost is achieved if only one insemination is used at eighty hours but then a 'clean up' bull needs to be used for repeat breeding.

Apart from a routine pregnancy check some two months after the end of mating, minimal management input should be required until turnout. Target weight gain after conception is only 0.5 kg per day, which should be achieved with a high-quality forage diet.

Second Grazing Season

During the second grazing season weight gains of 0.7 kg per day should also be obtained without difficulty if good grassland management is practised. As the weeks to calving approach, a watch on body condition is required aiming at a level of 3.0 to 3.5. Over-feeding at this time has to be avoided otherwise large calves will be the result and there will be considerable dystokia problems.

Frequent inspection is required at this time, especially to watch for any signs of summer mastitis. Offering a small quantity of concentrates at each inspection helps the stockman to be able to handle the heifers and is especially useful when checking for the onset of calving. If the down-calvers are located near the milking facilities it is advantageous to allow the group to walk periodically through the parlour to get used to the new environment. After one or two visits the pump and pulsators can be switched on so that when a heifer calves and has to be milked she is less likely to cause problems in the parlour.

Replacement heifers are the future members of the herd so that it is well worth while giving care and attention not only during the rearing period but especially at calving time.

Chapter 11

BREEDING, AI AND BULLS

A MAJOR objective in profitable dairying is to get 85–90 per cent of the herd in calf each season with a calving interval of no more than 370 days. Survey results show that apart from very old animals it is failure to conceive that is the most important reason for culling dairy cows of any age. Suckler beef cows also need to calve regularly in order to maintain a tight calving pattern and so ease feeding and management problems. Good herd feeding, housing, freedom from disease and bull selection are factors which contribute to successful breeding but there is no doubt that one of the key factors is stockmanship. It is generally agreed that only a small percentage of cow fertility is influenced by heredity, the remainder being in the hands of the management and herdsmen.

Regular calving increases the output of milk and calves, it also reduces veterinary costs and the loss of good cows through slaughter. As the levels of feeding and management in a herd improve it becomes increasingly important to have animals with a high genetic potential, so that the selection of sires is another subject in which herdsmen should be closely involved. Breeding is a long-term process with no guaranteed results, although one hopes that by careful selection of bulls, heifers will turn out to be better cows than their dams. However, this will depend on how well the genes of the sire and dam 'nick' together. There is little doubt that in the future, computer programmes will increasingly be used as a method of sire selection. Some essential data such as milk yield and quality of progeny will be automatically gathered but herdsmen will continue to have an important role in providing information about such factors as cow temperament and ease of milking.

This chapter deals with the role of the herdsman in managing fertility. It first deals with the female aspects of reproduction, then with AI, DIY AI, ova transplants and concludes with bull-rearing and management.

134

Infertility

Some infertility problems arise from infections of the reproductive tract of the cow. These are usually localised non-specific types of infection although others such as venereal disease are spread by natural service. Brucellosis was, until its recent eradication, a major infection of the genital tract which seriously affected fertility. Animals which have difficult calvings, torn or bruised tissue or a retained placenta commonly develop infection of one or more sections of the breeding tract. These animals produce abnormal discharges which are foul smelling and because the pus is a browny or yellowy or white colour, it is often known as 'whites'. This problem requires quick spotting and prompt treatment by the vet. Correct diagnosis is another important skill required of the herdsman. 'Whites' should not be confused with the slightly cloudy often blood-tinged discharge which sometimes appears two or three days after oestrus. This is quite normal and even if the cow has been served, it does not mean that the cow has failed to conceive.

Post-calving Management

A cow which calves and cleanses normally and is well fed in early lactation, should come into oestrus within three or four weeks after calving (which needs to be recorded). The uterus is stimulated back into normal condition by oestrus, due to an increased flow of blood in this region of the body. Cows not seen in oestrus by forty days should be examined by the vet. The length of gestation is on average 280 days so that to produce a calf at exactly a one-year interval, the animal would need to conceive on the 85th day of lactation. A useful guide in practice is to serve the cow at the next oestrus after fifty days in lactation, assuming she has already had a previous oestrus, because attempting to serve at the first oestrus reduces the chance of conception. Following a difficult calving or retained placenta it is preferable to delay service to allow time for full treatment of any infection.

Research has shown that the chance of conception is improved if the cow at service has started to regain the weight she will usually have lost in the early weeks of lactation. Obtaining accurate body weight is impractical on most farms so that condition scoring as described earlier is a practical alternative, the 'guide' score at service being 2.0.

Other research has shown improved conception from cows well fed, especially in energy terms, at the time of service. The diet needs

70. A rotary breeding board used to predict forthcoming events such as calving, oestrus or drying off.

71. The cow standing to be mounted is in oestrus and ready to be served or inseminated.

to provide adequate energy for maintenance, milk production and 10–20 extra MJ for liveweight gain.

Breeding Records

Chapter 14 deals with the collection, analysis and use of records, including those required for herd breeding. It is however worth stressing at this point that an efficient recording system is a key factor in successful breeding. In beef herds using natural service the only data required are the dates when the bulls were turned in and taken out but with a dairy herd a comprehensive system of recording is justified.

Heat Detection

In herds which rely solely on natural service there is less need for herdsmen to be fully skilled in heat detection. However, even in herds where cows have to be taken to the bull, time can be saved by only presenting the ones that are well in heat and will stand to be served. In the majority of herds, i.e. those using AI, it is a vital part of successful breeding management that herdsmen, and in fact all farm staff, are skilled and effective in heat detection. It is not a simple task, as studies have shown a very wide range in the length of time that individual animals are in oestrus—mature cows from only half an hour to twenty-five hours, with maiden heifers seldom for more than twelve hours. A few animals will understandably be missed if the oestrus occurs during the night hours, but the aim should be to identify 85 per cent of predicted heats (a simple calculation which can be undertaken weekly or even monthly). During the breeding season, three, or even four periods of at least half an hour need to be set aside solely for this purpose. These sessions should preferably be at times when the herd is settled and is not expecting to be fed, moved or milked. One of the key times is late at night so that adequate lighting should be available in winter housing.

A cow is only right for immediate insemination when she stands to be mounted. In herds practising batch calving, or in large herds where cows are yarded by calving dates, several animals may be in oestrus at the same time and form a group, making accurate identification most challenging. It is also necessary for stockmen to recognise the behaviour of animals coming into and going off heat so as to arrange appropriate timing of insemination. In the majority of situations, the inseminator will come to the farm following a phone

72. Purpose built AI stalls adjacent to the exit from the milking parlour.

73. Artificial insemination.

call, at approximately the same time each day. It is necessary, therefore, to have an appropriate cut-off time, after which any cow seen coming into heat is saved for the following day. When 'Do it Yourself AI' (DIY AI) is in operation, there is the opportunity for insemination to be carried out at the optimum time, i.e. when the cow is standing for mounting, although in practice the job is usually fitted in as and when convenient.

Numerous heat detection aids are available to help identify oestrus in problem cows. These include the Kamar heat mount detector which is stuck with glue on the sacrum of the cow, just in front of the tail head, changing colour when she is mounted by

| Poor | Fair | Good | EXCELLENT TIME TO BREED | Good | Egg re-lease from ovary |

Coming Into Heat	True Heat	After Heat
1. Smells other cows. 2. Attempts to ride other cows. 3. Vulva moist, red and slightly swollen.	1. Stands to be ridden. 2. Bawls frequently. 3. Nervous and excitable. 4. Rides other cows. 5. Clear mucous discharge (late in period.)	1. Will not stand to be ridden. 2. Smells other cows. 3. Clear mucous discharge from vulva.

Source: MMB.

Figure 12. The best time for breeding.

another cow. A more recent development is a paint-like paste which is put on the cow just in front of the tail head, and which is rubbed off when the cow is mounted. It appears to last for some two weeks before it requires topping-up.

In some circumstances, as for example with heifers wintering in an 'away' location, it may be preferable to use 'controlled breeding' drugs (prostaglandins) to synchronise heat within the group. This enables the insemination to be undertaken at pre-arranged times without the need to identify oestrus (two inseminations being usually carried out on successive days). It should be remembered that these products are not 'fertility drugs' and good conception results will only be achieved if the cattle are cycling normally, receiving an adequate diet and in a suitable condition for breeding. 'Sweeper' bulls are normally used to serve those animals which do not hold to the inseminations.

Another valuable aid to good conception is to ensure that the animals are securely and safely restrained when presented for AI. It is usually convenient for the inseminator to 'operate' on cows tied in a cowshed, retained in a crush or yoke or even in an abreast parlour, but not so convenient in a herringbone. Many dairy layouts incorporate a set of purpose-built stalls into the exit from the parlour which are ideal for AI as well as pregnancy diagnosis and many other veterinary tasks.

If no one is to be around when the inseminator calls it is essential to identify each animal clearly by tying a label (supplies available from the local centre) or chalking the details on a board above the cow. It is also necessary to leave soap, water and a clean towel for the inseminator to wash down and carry out his routine disinfection, in order to prevent the spread of disease. A bin should also be available for him to dispose of used straws and soiled plastic gloves. It pays at all times to co-operate fully with the inseminators, so that they look forward to coming to the dairy and to helping achieve good conception results.

DIY AI

A number of owners, particularly those with larger herds, have purchased semen storage flasks for the farm and have arranged for members of their staff to undergo training in insemination. (At least two people need to be trained to cover for sickness and holiday periods.) Some saving in cost is claimed and it also creates additional interests for well-motivated herdsmen. It is, of course, advantageous in severe weather conditions or in a foot-and-mouth outbreak when the inseminators cannot get to the farm. As mentioned above, it is possible to inseminate at the optimum time and even to double-serve in certain cases. Conception rates are variable and depend considerably on the skills of the individuals doing the job, with retraining sessions being a necessity before each breeding season. There is considerable extra paperwork involved and regulations to be followed, including the provision of a locked container for the storage flask.

Pregnancy Diagnosis (PD)

It is advantageous to arrange for the vet to carry out pregnancy diagnosis as soon as possible after service. For many this will be after

74. Withdrawing semen from an on-farm storage flask.

75. Undertaking the paperwork involved with DIY AI.

some 90 days, although a number of highly experienced practitioners can make accurate assessment at 40–50 days. The MMB offers a test which measures the level of progesterone in milk to indicate pregnancy. A sample of milk is taken from the cow between the 18th and 24th day following service, the result being received within several days. Accuracy is over 90 per cent for negative results and over 80 per cent for positives.

Ova Transplants

This is a relatively new technique for commercial farmers but due to cost, tends to be restricted in use to very high-value animals of both dairy and beef breeds. The donor cow is transferred to a specialised centre for the 'operation' to recover fertilised eggs and implantation into the recipient cows or heifers.

Selecting Sires

This is a challenging task. Nevertheless undertaking or being involved with the management in the selection of sires to use both on cows and maiden heifers can be a most rewarding aspect of a herdsman's job. Advice on this subject is available from a wide range of sources but those who have to take decisions are strongly recommended to contact their AI Centre Manager or ADAS Livestock Advisers for the latest information. A wide range of bulls of many breeds are readily available from the MMB Independent Cattle Breeding Centres and Breeding Companies. The bulls are of varying potential, proven ability and, of course, price!

In a 'flying' herd where all replacements are purchased, a beef sire which produces high-value calves, but without undue risk of calving problems, will produce better results. In a stable-sized dairy herd rearing replacements, it may be preferable to breed the best 60 per cent of the herd to top proven sires and the remainder to beef bulls. In selecting proven sires, the production figures of his progeny are of major consideration but so is confirmation, reliability of the data, his fertility and the value for money of the semen.

In order to evaluate genetic merit, the AI organisations have developed sophisticated techniques using Improved Contemporary Comparisons (ICCs). To eliminate the influence of feeding and management which account for much of the difference in performance between herds, daughters of a bull are compared with daughters of other bulls calving in the same herd at the same time. Lactation records are corrected to allow for the age and month of calving of the heifer. Only those calving in the same four-month period are classed as contemporaries. An ICC is the average amount that a bull would be expected to transmit to his progeny related to a bull ranked zero. In selecting a sire, as long as he has higher genetic merit than the cow he is to be bred to, improvement in performance is to be hoped for. It is preferable to select for weight of fat and protein and not for percentages. The highest income is obtained from animals which produce the greatest weight of butter fat and SNF. If the herd figures for fat are near the penalty zone, then in that case it would be wise to select a sire with high percentage fat figures. Included in each set of data for a bull is the 'weighting' which is a figure indicating the reliability of the ICC. As more daughters complete lactations, the weighting increases and only bulls with a weighting over 30 can be used with confidence. It is preferable for the daughters to have been milked in at least ten herds with no more than 25 per cent in any two herds.

Conformation ratings are provided by each of the numerous breeding organisations, often on a differing basis, so that valid comparisons can only be made between bulls evaluated by the same scheme, i.e. MMB or British Holstein Society. The main conformation factors are udder shape and attachment, teat shape and placement as well as legs and feet. Most organisations provide a useful brief summary of the typical daughters of a bull, e.g. large cattle, Canadian-type, with good udders, legs and feet just above average.

Rearing Bulls

It has been common practice to single-suckle bull calves being reared for service and in some beef breeds even to foster the calf on to a dairy cow. Bucket rearing, however, provides an entirely satisfactory start for a bull calf but it may, with advantage, be weaned a few weeks later than a steer or heifer (at 7–8 weeks). A bull should be handled frequently from a young age, taught to lead on a halter and be spoken to, so that his confidence is gained. A ring needs to be fitted by nine months of age, usually using a self-piercing ring into the septum of the nose, but it is preferable to bore a hole with a punch so as to locate the ring more accurately. When handling an older bull it is essential to use a staff and to keep his head held up high. After sexual maturity, at 9–10 months of age, it is necessary to pen bulls separately but also to maintain regular exercise and handling on hard ground to help keep the feet in good shape.

76. A cattle progeny-test panel involved in a 'workshop' session to evaluate a new method of classifying conformation.

77. Handling a difficult bull on two poles.

Housing for Bulls

With older bulls, especially of the dairy breeds, it is essential to have properly designed bull boxes with adjacent exercise yards. The internal walls need to be smooth to avoid damage to the bull and a feeding hatch incorporated, to make feeding easy and safe. Attached to the exercise yard should be a service crate where the cow to be served can be let in through a gate which shuts off the crate from the bull yard. In order that the stockman can clean out the pen in safety, either a heavy sliding door can be used to separate the box from the yard, or a feeding stall can be installed with a guillotine gate at the back to hold in the bull. It is necessary to be able to catch the bull; this can be done at the manger if a yoke is fitted (not really practical with a horned animal) or by having a short length of fine rope fitted to the ring to catch him while he is feeding.

Feeding the Bull

The overall aim is to keep him fit without being fat. Many young beef bulls being prepared for sale are allowed to get too heavy and may have to be slimmed down before working effectively. Unless the bull is working hard a forage-based diet may be adequate, hay usually being most convenient. 1.3 kg per 100 kg liveweight is a guide with 2 to 3 kg concentrates per day when in active use. During the summer it may be possible to tether the bull on pasture. This system affords exercise and avoids boredom, but he will need some shelter from the hot sun.

Working the Bull

A young bull should not be overworked in the early part of his life. He can be turned into a yard with a small group of heifers to play around and arouse his sexual desires. Once he has successfully served a few heifers, he can then be used on bigger cows. If a very big animal is to be served it may be necessary to dig a pit for her hind-legs to stand in. At all times slippery floors should be avoided. Some older bulls become slow in service owing to lameness in a hind foot or fear of pain or injury, or even strange surroundings. It may help him to be turned into a safe yard with a group of cows to build up his confidence. During the 'off-season', loneliness and confinement often lead to a deterioration of temper and when the new breeding season commences he may need to serve a cow more than once as the semen will initially be of low fertility. When a beef bull is

78. A Bull pen located alongside the cow yard. This is a valuable aid to oestrus detection as well as being convenient when natural service is undertaken.

79. A vasectomised bull fitted with a chinball marker. This is another useful aid to oestrus detection.

run with the herd he can easily be overworked, so it may be necessary to have at least two bulls so that they can be switched around and each allowed a rest to freshen up. In high-yielding herds where concentrates are fed with other feeds out of the parlour, there is a danger that bulls which run with the herd will get too fat. Not only does their work suffer but over-weight bulls can damage parlour gates and stalls if allowed in with the cows. A bull which is producing much semen and showing excessive signs of libido is not necessarily fertile. If it is found that cows are not holding, then the vet should be asked to take a semen sample for quality testing. When buying a bull, it is advantageous to have his semen quality tested before using him in the herd. It is wise to obtain a guarantee of fertility, use him on a few heifers then have a PD carried out as soon as possible.

Another reason for having a bull on the farm is to help with oestrus detection for AI. Locating the bull in a pen alongside the cow yard or adjacent to the exit route from the parlour attracts the cows in oestrus. Some farms use vasectomised bulls (teasers) to aid oestrus detection, usually fitting them with a chinball marking device on a head-collar. When mouting a cow, he leaves a paint mark on her back; colours are usually changed every three weeks. The same problem can arise as with an entire bull when many cows come on to heat together, and one is missed being marked.

Bull Health

One of the main difficulties with natural service is the risk of transmitting infection within the herd. Genital vibriosis is spread in this way and can be a cause of abortion at four to five months. There is a particular risk with hiring or borrowing a bull so that treatment for possible vibriosis should first be arranged. It is important to keep the feet of a bull in good shape and it may be necessary to cast him or have cattle stocks available in order to carry out foot trimming.

Lice can also be a problem in winter and it is worth applying louse powder or other pesticide every month and clipping out the top of his neck, withers and along the back.

It can be seen that for a successful breeding programme to operate, considerable time, effort and skill is required from herdsmen. As herd health and nutrition are closely involved in fertility, team work is required so herdsmen need to work effectively with veterinary surgeons and feed advisers as well as with herd owners.

Chapter 12

PREPARING STOCK FOR SHOWING AND SALE

MANY HERDSMEN will not be involved in showing or in the need to prepare cattle for sale as the majority of animals culled from the beef or dairy farm go directly to slaughter. Some, however, will be involved in this activity and others may have ambitions to work at some later date with a show herd. Pedigree breeders who participate in shows are provided with a shop window to exhibit the type of animals they may have for sale during the year or in the future. Showing also provides a good opportunity for herdsmen to obtain additional job satisfaction, by demonstrating their skills in the preparation and in the showing of the animals. Producers wishing to purchase breeding stock should primarily be concerned with production records but, as explained in Chapter 11, the body conformation and appearance of stock is also an important consideration. This chapter is written for those doing the showing rather than for those wishing to buy stock. However, the latter group may wish to read on—despite the fact that it is not the intention to declare too many 'tricks of the trade'!

The first and most important factor in preparing cattle for show or sale is that the job must always be done to a high standard. It is a poor advertisement for a herd if the cattle are only half-washed or badly clipped. Such animals in a sale could well make a higher price by not being touched. This is another example of a job which, if it is to be done effectively, must be done by a skilled and experienced person who is allowed adequate time and has the required equipment to hand.

Selection of Stock

The process of successful showing starts with effective planning. It is first necessary to obtain the date of the event (perhaps even more than a year in advance) so that action can be taken to have the cattle at an appropriate productive stage for the various classes, e.g. cow in milk. It is also important when, for example, one is selling surplus in-calf heifers, to locate a sale date to coincide with the cattle being at their best, i.e. a few weeks pre-calving.

The animals one intends to enter for a show may have to be selected out from the herd several months before the event so as to enable feeding and management to produce ideal body condition. When halter training is involved this cannot be undertaken with confidence in a short period of time.

Feeding

The feeding of show cattle is an important aspect of preparation, and on many farms involves diets somewhat different from those fed to the commercial herd. The main objective is to produce a good bloom to the coat and an adequate level of fleshing (condition scores not usually quoted!). The forage:concentrate ratio will usually be similar to normal diets but it is the constituents of the diet that tend to be different. Bran and sugar beet pulp are ideal foods, helping to stimulate appetite and keep the gut in good function. Oil-rich feeds such as linseed meal are frequently used to help produce the required bright skin and hair. High-quality hay is the most suitable background feed and is convenient to handle, especially at the show. A number of good bales should therefore be put on one side specifically for the show cows.

Managing Show Cows

As most of the shows take place in the summer months animals brought into the buildings for preparation need a cool location with good ventilation. Adequate supplies of clean bedding are obviously required, sawdust being a very useful material. Cattle being prepared for sale during the winter months benefit from being yarded in a separate group to allow appropriate feeding to take place.

Regular exercise is a necessity for show animals, so that time needs to be available to walk the animals, especially a bull, around the farmyard. Even when exercising, it is preferable that a leather rather than a rope halter is used, to avoid rubbing off any hair. The exercise period is a good time to teach the animals to stand well. It involves keeping the fore-legs apart and one back leg slightly behind the other. The head should always be held up.

80. Clipping the udder.

81. Washing for the show ring.

Clipping

This is a very important part of show preparation and needs to be carried out with skill and care. Clipping will need to be undertaken some time before the show or sale, with some additional work usually needed a few days before the event. Electrically-driven clippers are most convenient; there is a wide range of cow blades and plates available, including special fine ones for show work. A good grooming before starting the clip prevents the blades losing their edge. Special aerosol sprays are available which, as well as lubricating the blades, disinfect them and also blow out the dirt and hair.

The areas for clipping include the head, neck, shoulders and udder, as well as the tail and topline. One of the skills is to use the technique to minimise the appearance of defects, so that along the topline, for example, close clipping of high points and shorter clipping of lower points really does improve appearance. The amount of hair removed at the front or rear of the udder will depend upon the quality of attachment and prominence of the milk vein. Care should always be taken to ensure that the areas clipped blend in well with the areas not clipped, this being easier in summer when there is less hair to deal with. Clipping only the head, top of the shoulder and the tail of cattle for sale can markedly improve their appearance. It is often tempting to clip out the legs as well but this can be a rather dangerous activity when undertaken on young heifers and even on cows!

Washing

This is a routine part of preparation and for show cattle approaching the event will need to be done at least weekly. Not only does a good washing remove dirt and dead hair but it stimulates the skin and generally improves coat condition. Warm water and liquid detergent are ideal (although proprietary shampoos are available), taking care not to allow soap and water into the eyes and ears. Blow driers are frequently used after washing to give an improved appearance to the coat. Although stained skin will be rare in show cattle, it can be a problem when preparing animals for sale during the winter months, especially with cubicle refusers—which are the ones which need to be sold! It may be necessary to wash the stained areas on several occasions and then to use some chalk whitening on the day of the sale.

Trimming

The feet of show and sale cattle must be in good order. Any paring or cutting of the hoof which is required needs to take place well before the event to allow the animal time to adjust its standing and walking to the new angle of the toes. The horn of the foot looks better if scraped until smooth, followed by polishing with horn or shoe polish. The same technique can be used to trim and polish horns, especially of bulls.

Grooming

After thorough washing and allowing time for the coat to dry, grooming can take place, as ideally it should every day before a show. A stiff brush is best suited for the initial grooming, as is a rubber curry comb, followed by use of a softer brush in conjunction with a lightly oiled cloth, just before parading. The tail needs particular attention, as a well-groomed and fluffed-out tail certainly adds to appearance. The herdsman leading the animal should always be well 'groomed', wearing a clean and well-ironed white coat, polished shoes and have a general neat appearance. Leather show halters and leads also need to be regularly cleaned and polished, a special one being kept for use in the show ring.

Bagging-up In-milk Cows

It is obviously unwise to milk out a cow several hours before she is to enter the ring, but it is equally important that her udder is not too full, causing restlessness or the teats beginning to 'poke'. Knowing the time of parading, it is necessary, therefore, to milk out as appropriate, this being another example of the skill of the herdsman in knowing the precise requirements of each cow, and even of each quarter.

A sharp appetite needs to be maintained as the show approaches, with a good feed and drink being allowed just before parading takes place. The likes and dislikes of individual cows should also be known so that a 'little bit of what she fancies' can be offered at the appropriate time.

Parading

There is also some skill in walking the animal slowly but effectively in front of the judge or the buying audience. One needs to be calm

82. Blow-drying. This is a useful way to improve the appearance of beef cattle with long hair.

83. Fluffing out the tail.

but firm with the animals, sometimes walking backwards, so as to be in a position to check that the animal looks at her best. When asked to stand for close inspection by the judge, it is important to have the animal standing well, with head up and legs positioned as described above.

Appearance of Show Stand

Although the parade of the animals in the ring in front of the public and other breeders is the main shop window, much potential business is made or lost by the general appearance and tidiness of the exhibitor's stall area. The equipment and feed should be neatly positioned, keeping most of the tools in the show box. Someone should be 'on duty', especially at busy times, not only to keep an eye on the stock but also to answer questions and generally be helpful to potential customers. The availability of a herd brochure may well be justified; or if not, some performance data (neat and well presented) should be on display above the animals. Many show societies positively encourage such an attitude by awarding prizes to the best breeder's stand. Additional factors taken into account when judging these classes are cleanliness of lines, utilisation of space and overall effect for herd and breed.

Showing can be a pleasurable and satisfying task for the herdsman. It involves long hours, hard work and, on most farms, the work is piling up awaiting his return from the show.

Chapter 13

MAINTENANCE OF MACHINERY AND EQUIPMENT

IN THE PAST, the herdsman has taken little interest in farm machinery as his job was seen to be caring for stock and he was content to leave the caring of those 'mechanical gadgets' to others. Furthermore, farm people with a preference for driving tractors have tended to be located on arable holdings and have avoided too much involvement with farm livestock whenever possible. Occasionally, a herdsman may have been seen oiling the wheel of his barrow but only when the squeak was so serious as to disturb his cows!

Gradually, however, the scene may be becoming less clear cut. Mechanisation of cattle units has taken place slowly but progressively over recent decades, first, with the introduction of the milking machine, then feed and manure-handling equipment and now with automation and electronics. It is now fair to say that a number of today's stockmen do recognise the importance of the machinery and equipment and take a keen interest in its maintenance but there are still many others who do not. With the continuing trend towards increased herd size and a reducing labour force, however, the essential role of reliable machinery cannot be over-stressed. As with the cows and the veterinary surgeon, outside, specialist help is often required for major repair work but the need for this is reduced if the regular stockmen understand the operation of each machine they use and are involved in the routine servicing and adjustments. It is therefore the case that the new generation of herdsmen need to have a much wider range of skills at their fingertips although there are limits and it is not suggested that they should be able to write computer programmes or even replace the clutches of tractors! This chapter is intended to help the stockman recognise the value of efficiently maintained equipment by showing that it responds just as much to his care and attention as do his cattle. Just as with animals, careful observation of machinery is needed and action must be taken at the first signs of a fault.

One of the major problems facing a stockman is to find the time for all the numerous jobs that need to be done each day. Problems seldom appear at a steady rate so that they can be tackled there and then; more often than not one problem leads to another and then of course, the animals must come first. But unless machinery problems are given some regular attention a 'vicious circle' can be set up. Routine jobs being undertaken with inefficient machinery take longer than they should so that, in turn, there is less time for vital maintenance. One of the most common examples of this is the bent or damaged slurry scraper which continues to be used causing additional damage to doorways and even to cubicle uprights. If planning the use of one's time is important, even more so is making sure that those plans are carried out—and especially that the difficult or unpleasant jobs are not just left undone.

Herd owners and managers have a particular responsibility in this area making sure that they recognise signs of stress in herdsmen, avoiding their constant overloading, and arranging for appropriate back-up services. Fewer problems appear to exist on farms where owners and managers are actually involved in the routine tasks, perhaps by undertaking relief duties. This gives them a clear picture of what the job entails, how well the machinery operates and, for instance, whether there are the necessary tools to hand to cope with essential repairs. Small farms and one-man dairy units often tend to suffer most from a lack of adequate time to deal with machinery problems whereas the larger units usually have at least one member of the team of stockmen with the time and interest required.

Some dairy and beef units are part of much larger farming businesses often associated with arable enterprises so that a farm workshop is available, with staff to deal with the herdsman's equipment problems. Relationships between herdsmen and farm mechanics are not always good, due to their contrasting interests and to a failure of each to understand the others' point of view. Herdsmen can make a positive step to improving the relationship by, for example, making attempts to wash off as much slurry as possible from the damaged scraper before dumping it on the workshop floor. Inviting the mechanic to inspect a problem at an early stage can be far more rewarding to all concerned than waiting for a complete stoppage and then 'passing the buck' to the workshop. Similar attitudes are known to exist in enterprises which are sufficiently large to justify stand-by equipment. If stockmen abuse such a resource then they often find that both machines are out of commission and considerable inconvenience occurs.

Even on farms where there is a well-equipped workshop, it is

84. Washing-off the slurry scraper for a maintenance check.

85. Using a power washer to clean equipment.

advisable for the cattle unit to have a small set of essential tools for use when the workshop is closed. This should include such items as: hammer, pliers, screwdrivers, as well as adjustable spanners and any specialised tools required to repair drive chains. In situations where more than one person uses the tools there needs to be an established storage place where each item can be returned after use. The location would usually be in the dairy office or store, along with such things as calving ropes and medicines. A more comprehensive set of tools will be required on farms without a specialised workshop and these may be so valuable as to justify a lockable store.

The availability of spare parts is another important factor in efficient machinery operation. Carrying too many parts on the farm is expensive because it ties up cash, but carrying too few creates obvious problems with frequent journeys to and from the suppliers. When a new item of equipment is obtained the dealer's experience should be used to help decide which and how many spare parts should be held at the farm. A very satisfactory system for replacing used parts is to attach a label to each important part in the store recording: for which machine, part number and local supplier. When the part is used, the label is removed and placed in a box marked 'shopping orders'. When someone from the farm is next going to town, they take the labels for part numbers. On return, with the replacement, the same label is attached and the item put into store. A similar well-organised method of storing instruction manuals and parts books is recommended. A small metal cabinet with drawers is ideal and, again, can be located in the 'stores'.

On some farms, the 'boss' may prefer to keep such books in the farmhouse or general office, so in those cases, stockmen should be supplied with copies.

Mention has already been made of the arrival of new items; this is an important occasion as efficient operation is so often determined by the success of the 'induction ceremony'. All expected users should be present, including 'the Management', to have potential problems pointed out and to agree a maintenance programme. This may not be required for the infamous slurry scraper, but essential for such items as bulk tank condensers, forage boxes or electronic feeders. The name and telephone number of the service agent needs to be recorded in the dairy as well as in the farm office, along with the veterinary surgeon and AI Centre. It can be advantageous, especially with less common equipment such as a complete diet feeder, to have an arrangement with a neighbouring farm which operates a similar item, so that a stock of spares can be shared and information exchanged on solutions to potential problems.

86. A dairy tool box containing specially-sized spanners to fit the equipment used and maintained by herdsmen.

87. A slurry separator. A good example of a machine which if well designed, constructed and maintained, will reduce the time involved by herdsmen in dealing with slurry. If not, breakdowns lead to increased time involvement and frustration.

As with cattle breeding records it is a considerable help in the planning and carrying out of machinery maintenance to have some suitable method of recording. A small blackboard or calendar may be adequate, listing the machines and date of the next service. If anyone then has a spare hour during the week, they can check the list and find which item is in most need of their attention. Knowing how many stockmen operate such records, it may be easier for them to have a few specially coloured pins in the actual breeding board— not indicating when the old tractor is due to calve but when it should have an oil change!

A useful alternative is to use Action Plan Lists to remind one of the jobs to be done before each seasonal activity. Such lists can be obtained from machinery or fertiliser manufacturers, but can be easily prepared. One for use before the winter feeding period would include such items as:

Machinery
Slurry Scraper
☐ Check condition of frame, especially wings and rubber blade.
☐ Have spare blades ready.

Forage box
☐ Service according to instruction manual.
☐ Check universal joints, especially at drawbar.
☐ Check spare parts stock, especially drive chains.

All these 'on-farm' aids such as a good tool set and appropriate recording systems considerably help the efficiency of machinery maintenance. But another factor of major importance is the basic understanding of the subject by the person carrying out the task. Many younger stockmen now coming into farming have had the opportunity as part of a general agricultural course to study various aspects of mechanisation. Others may not have had such an opportunity, so it is to their benefit to identify and take part in specialised training events to improve their knowledge and understanding. The ATB and many County Colleges offer such courses and hopefully will increase this activity in the future.

Efficiently operated machines not only cost less in repairs but on the cattle unit they also contribute considerably to increased output. Stockmen who can carry out their routine duties effectively and without the worry of frequent breakdowns have more time for their key job which is to observe and care for their animals.

Chapter 14

RECORDS AND RECORDING

PAPERWORK IS not the most pleasurable part of the job for many herdsmen, although most do appreciate the need in a present-day enterprise for some records to be kept. In earlier years, generally with much smaller herds, minimal recording was necessary because the herdsman's memory, aided by a few dates marked on the cowshed wall, served the purpose well. The effective management of a cattle enterprise in the current economic climate, however, requires the collection, analysis and subsequent use of considerable quantities of data. The role of an individual herdsman in this task will vary widely from farm to farm but whenever it is undertaken it is another important task requiring once again care and attention to detail. Accuracy is most essential as incorrect information is not only a waste of time and paper but can create extra cost and loss of output.

The advantages of an effective system of records should be obvious to the herdsman as much of the data recorded is subsequently used by him to carry out routine tasks such as drying off or diverting cows for pregnancy diagnosis. Other data he may be expected to record such as feed usage and fertiliser application is required for management purposes with which a good herdsman should be concerned.

Some records such as livestock movements have to be kept by law, but others are entirely optional. One problem is that there is such a wide range of recording systems available, especially for dairy enterprises. Time can be wasted and confusion is frequently caused if several systems are in operation, involving some duplication of recording. There is a considerable responsibility for herd owners and managers to set up a simple but effective system for herdsmen to operate. Everyone working with cattle will at some time be involved in record keeping if only noting the number of a cow seen in oestrus; the actual level depending upon responsibility within the enterprise. During weekends or holiday periods such responsibility may be

161

temporarily increased so that a clear understanding of what is involved should be communicated to relief stockmen. A serious problem can arise if essential data is not recorded, as it can if the head stockman's books are incorrectly completed by an over-keen farm student!

Following collection, much of the data has to be analysed (worked upon) before it can be used in decision making, e.g. in culling. The involvement of the herdsman in this analysis will also vary from farm to farm depending largely upon the back-up service provided by the owner, farm office or off-farm organisation. Some of the events, such as calvings, will be written down and passed 'down the line' but they will also be entered on the herdsman's visual display board.

In the few very modern enterprises, data such as individual milk yields is automatically collected and fed into the memory of the on-farm computer. The data is processed immediately so that any cow required for special treatment, e.g. antibiotic milk to be dumped, is identified by a flashing light on her stall. The record-collecting role of the herdsmen on such farms is very much simplified so long as the equipment is reliable.

This chapter is written to aid herdsmen undertake the basic task of keeping and using records, rather than attempting to describe the various alternative systems that are used within cattle enterprises. Advice on the selection of an appropriate system for a particular enterprise is readily available from such organisations as ADAS, MLC and MMB. Several feedstuffs and fertiliser manufacturing companies, as well as offering enterprise costing and monitoring schemes, also supply useful cards and charts on which herdsmen can record raw data.

As an aid to accuracy, timely recording is essential. Events need to be recorded as soon as possible after they take place by being written into a book or on to a card, chalked on to a blackboard or even punched into a computer keyboard. Delays lead to memory failures and time wasted, for example going back to check on the number of a cow seen in oestrus. Keeping a small notebook and pencil always on hand avoids this sort of problem. The ideal type is one which easily fits into an overall pocket and one which retains the pencil within the spine. Children's magic slates are used on some farms with herdsmen effectively scratching information on to the slate with a finger nail! A period of time needs to be built into the daily routine so that essential events are recorded on time. This is particularly important at busy times such as the peak of calvings when, although the herdsman is busy with the cows, even more data has to be recorded.

Although the importance of good records has been stressed, there is always the need for a balanced approach to the herdsman's job, and of course the animals should come first. It has been known for some herdsmen (as well as managers) to be so preoccupied with records that they spend excessive periods of time in the office, to the neglect of their stock or more commonly to the appropriate support of their junior colleagues.

Although, as mentioned, not all recording systems will be described, there now follows an explanation of a few key records in which most herdsmen will usually be involved.

Livestock Inventory

It is essential at all times to know the number, class and location of every animal in the enterprise. Such information is required at regular intervals for completing the MAFF returns but is of valuable general use in the day-to-day management of the herd. A simple monthly chart with a line for a day and a column for each class of stock is appropriate.

	Livestock Inventory			January 1983			
	Cows in milk	Dry cows	Incalf heifers	Yearlings	Heifer calves	Bull calves	Notes
1	95	10	12	25	17	0	
2	96	10	11	25	18	0	Heifer 437 calved H. Calf
3	96	10	17	19	18	0	6 Heifers PD + ve

In a large dairy herd which is subdivided into groups it is helpful to maintain a record of the location of every animal so that time is not wasted when an individual needs to be checked or has to be diverted for any purpose. A wall-mounted chart serves this purpose with animals listed in numerical order and with different coloured pins for each group stuck into the board, adjacent to each number.

A similar display can be used to locate groups of grazing animals by sticking large-headed pins into the appropriate fields on a wall-mounted farm map. The number of animals in each group can be

Herd Grouping						
Red = High yield (Yard 4) Blue = Medium (Yard 3) Green = Low yield (Yard 2) Yellow = Dry						
1	R	156	Y			
2	B	157	G			
3	R	158	SOLD			
4	G	159	G			
5	Y					

written on the head of the pin, a system ideal for assisting relief staff who are involved in checking outlying groups.

Outputs (Sales)

On a dairy enterprise herdsmen need to retain the daily milk slips left by the tanker driver. These are handed in to the office where they are checked against the monthly payment statement. Usage of milk by calves, staff and other outlets (retail sales) may also need to be recorded and a special monthly sheet prepared.

Cattle dispatched from the farm for sale, such as calves, culls or beef animals, should be accompanied by a dispatch note, especially when an 'outside' haulier is employed. A triplicate pad is ideal for this purpose allowing the top copy to go to the purchaser with the consignment, second copy to the office for checking against receipts, and the third copy retained in the book with the herdsman. Details should include date, numbers, identification, age and sex, as well as the purchaser. Although it is never a pleasant task to record deaths or sales to knacker's yard, it is essential for this to be done if the books are to balance at the end of the year.

Inputs (Purchases)

Herdsmen should be responsible for checking deliveries to their enterprises, such as feed, fuel or spare parts. Before signing delivery notes, quantities and other necessary details should be checked and any errors recorded. Delivery notes should then be retained, perhaps in a clip on the dairy wall, until forwarded to the office. Incoming sacks or containers which are not on pallets should be stacked neatly, so that subsequently they can be readily counted, perhaps for the end-of-month or annual valuation. Bulk containers (concentrate hoppers, fuel or molasses tanks) should ideally be calibrated, to help in valuation but also to enable stock usage to be more readily monitored. It is also of considerable help if hay and straw barns, as well as silage clamps, hold 'known' quantities.

Breeding and Fertility

As explained in Chapter 11, the need for herdsmen to operate an effective system of records is essential to a successful breeding programme. The use of a desk diary may be all that is required in a small dairy or beef enterprise to record daily events such as calvings and services. With only a few animals involved it is not difficult to look back, for example, over a three-week period to check on a previous service. Using a set of breeding tables, forward events such as drying-off dates can be calculated for entering in the appropriate dates of the diary.

In larger herds, in order to effectively monitor events a more comprehensive system of breeding records is justified involving, as well as written documents, some type of visual aid board. Various types of rotary and oblong boards are available which utilise coloured pins or markers to record specific events. The pointer or marker is moved forward each day, highlighting expected or planned events such as cows due for pregnancy diagnosis. Detailed operating instructions are normally provided by the manufacturers but it is essential for everyone involved, especially relief staff, to fully understand the system. In very large herds two or more such boards may be required to handle all the cows, which involves a considerable input of time in keeping the information updated. It is for this reason that other systems such as a card index system carrying event markers have been developed. A most satisfactory card was prepared by a commercial company for the 350-cow Sonning herd of the University of Reading. Before computerisation this was used for several years to record in writing the breeding management events on individual cow cards which were appropriately spaced in racks in a cabinet. Small coloured metal slip-on tags are placed on the right-hand edge of each card which is divided into weekly divisions and used to pinpoint key events, as with a visual aid wall board.

Health Records

All health problems should be recorded as an aid to the management of feeding, breeding and culling, as well as improved herd health. A diary may again be adequate in a small herd but individual cow cards are better for larger herds. The record should start with the calf and

88. A rotary breeding board. In this example, each cow is represented by a numbered magnetic cube.

89. Oblong breeding boards. In this large herd two boards are required to accommodate all the data.

90. A card index system being used to record breeding and herd health data.
Small coloured metal slip-on tags can be placed on the edge of each card to
pinpoint key events for that particular cow.

be maintained throughout the lifetime of the cow. This is an area in
which computerised recording can drastically reduce the time
involved in analysing health data, but not in recording the original
data.

Mastitis records are particularly valuable as an aid to culling the
problem cows. Original data is best recorded on a parlour blackboard then transferred to individual cow cards placed in numerical
order in a loose leaf booklet. The information recorded would be:

Cow No:.....................................

Date	Quarter(s)	Tubes	Sample Result

UNIVERSITY OF READING FARMS (SONNING FARM) OUT

COW NO. OPEN HERE

COW LIFE RECORD

RED SPOT
NOT TO BE SERVED

NAME

HB NO.
EARTAG NO.

SIRE

DAM NAME

GREEN SPOT
PD POSITIVE

LOCATION

GRAND SIRE

GRAND DAM

DATE OF BIRTH

DAM LINE NO.

AGE AT FIRST CALVING

TICK ✓
✓ in pencil

SUGGESTED BULL	(LACT NO.)	1	2	3	4	5	6	7	8	9	10	11

HIGH YIELD
LOW YIELD

EVENTS IN STRICT CHRONOLOGICAL ORDER

LACT NO.	EVENT	CODE	DATE	NOTES	LACT NO.	EVENT	CODE	DATE	NOTES

BUTTER FAT
LEGS
UDDER
MASTITIS
IN FERTILITY
WEIGHT
SCORE
MILK FEVER
TEMPER- AMENT
RUMEN FUNC- TION

JUL
AUG
SEP
OCT
NOV
DEC
JAN
FEB
MAR
APR
MAY
JUN

SIGNALS

BLACK = ACTUAL CALVING DATE

BLUE = SAFE TO SERVE

YELLOW = LATEST SERVE

GREEN = PD DUE

RED = DRY OFF

CARD NO.

Figure 13. Cow life record card.

Milk Recording

Official milk recording is organised in the UK by the Milk Marketing Boards. Milk recording can however be done privately but membership of an official scheme offers strict standards, a regular discipline and credibility of the results. The yield can be obtained by weighing in bucket plants, and in parlours using graduated jars, milk meters, or the new electronic weighers. For the official scheme, a recorder visits the farm on two (or three with 3 × milking) successive milkings each month to record yield and to take a sample for BF and SNF analysis. The recorder also takes details of calvings, services, drying-off and cows sold. He or she leaves a carbon copy of the yields, and the results, incorporating milk quality analysis, are normally received ten to fourteen days later. At the completion of each lactation a record card is received for each cow, detailing her lifetime performance and indicating her relative breeding merit within the herd, and nationally. An annual summary for the herd at the end of March details the performance of each animal, averages for the herd as well as the calving index.

Although involved in an official scheme, it is worthwhile undertaking more frequent recording, i.e. every week or two weeks, especially for cows in early lactation. This is a job well suited to the manager or owner who not only spends valuable time in the parlour monitoring the cows, but is able to talk with stockmen and discuss herd performance.

Monthly Costing Schemes

There are numerous schemes available to dairy farms for the budgeting, costing and monitoring of herd performance. Some, such as Milkminder, Checkmate and Herd Management Control, are available from the MMB Farm Management Services, but a wide range of others is available from commercial companies. The data to be recorded at the end of the month normally includes milk sales, cow numbers, feed and fertiliser usage. The statement returned several days later shows the calculated yield per cow, rate of concentrate use and margin over concentrates, total feeds or feed and fertiliser per cow and per hectare. Ideally this information can be compared with targets and action taken to correct deviations. As the information is by this time somewhat dated, it is better to keep a weekly or even daily eye on performance rather than wait for the monthly statement. With many schemes an adviser visits the farm either to collect the data or help with its interpretation. The latter role can be most valuable in using a 'new' pair of eyes to spot problems and discuss (with his experience in the district) the possible alternative actions.

Computerised Systems

Most of the above costing and monitoring schemes use computer facilities, when the data from the farm is normally punched into the system by a member of staff at the bureau or central office of the organisation. Some advisers now bring a mini-computer to the farm where it can 'stand-alone' or be linked by telephone back to a larger machine, so that data can be analysed immediately. Other schemes involve a group of farmers sharing a facility and perhaps employing a secretary/technician to process the data. A few larger farms have their own computer which offers herdsmen more frequent access to the information, should it be required.

With all computerised systems, accuracy of the input data is critical so that a simple but effective system of recording the raw data is essential. A duplicate book is ideal, as used by many farms involved with DAISY, the Dairy Information System of the University of Reading. Events are recorded as they happen, e.g. calvings, services, health problems, with the top copy going to the office on a daily or weekly basis for input into the computer programme. Output from the programme can be in the form of action lists for herdsmen (now on paper of easily handled size) or as displays on visual display units (VDU). No doubt in the near future such facilities will not only be provided in the farm office and milking parlour but in the hospital area for use by the vet and herdsman in health control, and even in the herdsman's home!

Chapter 15

CONCLUSIONS

THE OBJECT of this closing chapter is to summarise briefly the major points raised in the book, and, at the same time, to offer a few *final* thoughts to the reader. Some are indeed afterthoughts but others are general ones, concerning the herdsman's job as a whole, and have therefore intentionally been held over until this concluding chapter.

In the Introduction it was stated that with the continued trend towards mechanisation and automation in dairying, the days of the herdsman, as we have known him, are perhaps numbered. Having read the preceding chapters, herdsmen should have the feeling that nothing is farther from the truth and that, although the nature of their job may well change, they will continue to have a vital role in profitable milk and beef production. These days, change is accepted as a continuing and inescapable feature of the herdsman's job. In earlier years he had, primarily, to love or at least have a strong aptitude for dealing with cows; today he needs to also have an aptitude for handling machinery, for understanding paperwork and most importantly, for dealing with other people. It is against this backcloth of 'change' that the final thoughts in this book are presented.

The first *final* thought is simply that 'a successful cattle enterprise involves team effort'. The herdsman's fellow 'players' include, as well as the herd owner and other colleagues, the vet, AI inseminator, feed adviser and electronics engineer. A new tool of the trade is the herdsman's notebook which contains the telephone numbers of all his team members in case they are urgently needed 'on the pitch'.

Another important feature of the herdsman's job is its variety: requiring him to have a wide range of skills and giving him little opportunity to complain of monotony. Unlike the immigrant workers employed in the huge dairy units of the Western USA, contract milking on shifts, around the clock, generally speaking the UK

171

herdsman really has to be a stockman. Although routine tasks such as milking and feeding take up large parts of the day, there is much scope for planning and control, if efficient use of the remaining time is to be obtained. Another important *final* thought then, is that 'the management of the herdsman's time is a critical factor in the success of his enterprise'. In small herds, especially in one-man units, herdsmen have to manage their own time. Some are naturally better at this than others; with an opportunity for those who are not so good to correct this weakness. In large units, with several herdsmen employed, there is a considerable responsibility on the management to ensure effective planning of time and co-ordination of the effort of each team member.

On farms where there is poor use of time, there is a strong tendency for something, or someone, to be frequently neglected with detrimental effects upon profitability. It may even be the cows which suffer, or the herdsman's family, or perhaps most commonly, the care and maintenance of plant, equipment and buildings. A crisis can be expected from time to time on every livestock farm. In the enterprises where staff are well organised, this is 'taken in full stride' by herdsmen, putting in even extra effort, by working longer hours until normality returns. When crises frequently occur, this should be a clear indication that staff cannot cope and that the system needs some major change, which will inevitably involve the staff.

Frequent reference has been made in this book to the need for *timely* action to be taken by herdsmen in carrying out a wide range of their tasks, such as feeding, treating sick animals and keeping records. Although to achieve this, good organisation of time is involved, several other key factors are also involved. Here, therefore, are five general and basic points which, in this writer's opinion, can be crucial to the performance of the herdsman.

1. Adequate Knowledge. As the level of technology on the farm increases, so do the training needs of the staff. Learning therefore has to be an ongoing process, involving experienced herdsmen as well as new entrants. Despite the new technology, the basic skills of observation and caring for the animals remains the number one priority. The role of skilled staff in helping young herdsmen to gain sound experience cannot be over-emphasised.

An afterthought, here, from Chapter 8 on Herd Health: A valuable training session can be undertaken at the slaughter house when cattle from the herd are being culled, probably due to infertility problems. Arrange for the vet to be present as tutor, to demonstrate the working (and non-working) parts of the cow. You will then be able to discuss health problems with much better understanding.

Knowledge of the science and practice of feeding has been highlighted as a major factor in stockmanship, justifying three chapters of the book. A quotation from the 'King of cow feeding', Professor Bobby Boutflour, in the late '40s is still most appropriate to summarise the subject: 'Allow the cow to polish the manger with its tongue a dozen times a day.'

2. Adequate Motivation. Many herdsmen are happy in their work but others lack interest, feel isolated (psychologically as well as physically) and do not 'get on' with other members of the team. It is hoped that the herd owners and managers who read this book will have a better understanding of the role of their herdsmen and will try a little harder to give encouragement and more effective support, especially at busy times.

On their part, herdsmen should encourage effective communication, try to understand the owner's objectives for his farm and not take a job if not in full agreement with how the farm is expected to be run. An interview for a new position is a preliminary to forming a working partnership; candidates should take the opportunity to find out at the interview just what are the chances of that 'partnership' being successful. There are such contrasting employment situations within the industry, especially in terms of responsibility, that one needs to be entirely satisfied that the job being offered really does match the requirements.

3. Adequate Facilities. It has been emphasised in several chapters (milking, housing and machinery) that there is rarely any correlation between capital spending and financial performance of the enterprise. Good herdsmen appreciate good facilities but they also use their varied skills to 'make the most' of what is available. Employers, however, have a considerable responsibility if they wish to increase the efficiency of labour use by providing functional buildings and reliable equipment at a cost the enterprise can afford. Some facilities, although a minor cost on the business, can be highly beneficial in improving herdsmen's morale, by demonstrating that management cares. The provision of suitable protective clothing is such an example; it serves its purpose in protecting the herdsman but reduces the work of his family in washing cow-splattered clothing.

4. Home Support. If the herdsman is to give of his best to the farm it is advantageous for him to have support and understanding at home.

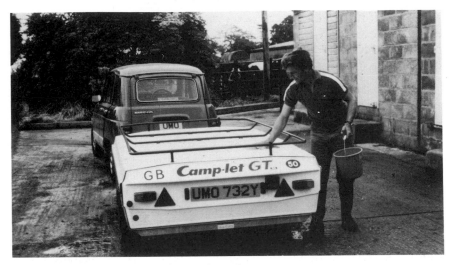

91. Preparing for a weekend apart from the cows.

The provision of meals when he is ready, as well as clean, dry clothing, are obvious, but after a hard day's work the comforts of a home rather than a house are invaluable. The well-known saying can be adapted to: 'behind every good herdsman is a supporting wife, mother or landlady'. The practice of young people living-in with the farm family is becoming rare, as is the availability of suitable digs in rural areas. The author is far from keen on what seems to be an increasing practice of single herdsmen living (existing) in sparsely furnished farm cottages with little time to prepare adequate meals, let alone undertake essential shopping. Unfortunately, many of the attempts to set up and operate hostels for young people have not been successful, primarily due to cost. So perhaps one recipe for a successful herdsman is to find a good wife; or a husband for a herdswoman!

5. *A Life Apart from the Cows.* Too many herdsmen (including small dairy farmers) are 'tied to the cow's tail'. Everyone working with cattle needs, and deserves, time off. As mentioned earlier, a number are keen gardeners, obtaining relaxation, despite remaining in the sight and sound of the herd. Interest in the job can be markedly improved by visiting other herds or shows (employers please note!), and also by having time away for hobbies and other outside interests. The younger generation of herdsmen do tend to be more mobile and have other interests, so in time this problem may well disappear. Senior herdsmen therefore need to learn the skills of delegation.

Relief staff respond to increased responsibility, so long as they know exactly what is expected of them. Effective communications and helping juniors to learn from their mistakes are the essential elements of delegation.

A Final Conclusion

If one message should have been clear from this book it is quite simply this: that a modern herdsman, with few exceptions, is a very fortunate person. He has a job in which he can take pride and from which he can obtain enormous satisfaction. His efforts, abilities and skills are reflected in the appearance and performance of the herd, which is measurable in the bulk tank, as well as in the herd health and breeding records.

The future promises much; some of his past and present-day skills (such as oestrus detection) will no doubt become redundant as automation develops further. Sex determination of calves will speed up the effects of a breeding programme. Development such as this—and many others—will be of increasing value as reflecting an electronic age; they will assume increasing importance as the needs of consumers change, and those changes become reflected in the demand for milk and dairy products—and for changing breeding criteria.

No doubt by the time the newest of *today's developments* are perfected, many of *today's herdsmen* will be enjoying their well-earned retirement—with a life permanently apart from cows! The majority, if not all of them, will be able to look back with satisfaction.

INDEX